Praise for Danny Iny's AI Methodology

A small selection of feedback from participants in Danny Iny's AI Strategist training, where the methodology in this book was first shared:

"Danny's AI training showed me how to leverage AI as a tool to maximize the most valuable asset in my business: ME! Learning to use AI to help me find clarity in a sea of confusion and bring order to a world of chaos, all while challenging my assumptions, has been a great learning experience." –JOHN MITCHELL

"I am a serious AI user and even train other business owners and professionals to use generative AI and have been doing so for more than a year. Nevertheless, Danny's training was eye-opening. The prompts he gave us and the perspective he shared was truly unique and provided me with key insights which will inform my work and my business as we move into this ever-changing future." –DR. STEVEN KIRCH

"I learned how to use AI as a collaborator and a tool to bounce ideas off with a focus on MY expertise, not random AI garble. I learned how a prompt filled with context and guardrails can help me create in a structured way until implementation. I highly recommend learning about AI from Danny, his take is different from what's 'out there' and the focus is on YOU not the AI tool." –ANNIKA EK

"This training was extremely useful, especially in contrast to most other AI trainings I've signed up for. This wasn't just a bunch of prompts but a real strategy for using AI effectively without falling into the common traps of AI."

—SOPHIE LECHNER

"Danny Iny has an extraordinary gift for taking something as complex and overwhelming as AI and making it clear, practical, and actionable. He provides frameworks, insights, and real-world applications that immediately spark new ways of thinking . . . Instead of being stuck in theory or hype, Danny showed me exactly how to integrate AI into my workflow in ways that save time, spark creativity, and open up new opportunities. I walked away not only with skills, but with the confidence to use them effectively. If you want clarity in the chaos of AI and a strategic edge for the future, Danny's work is a must."

—RAY ENGAN

"This training showed me how AI can release my own specific genius in a way that is consistent with who I am and who I want to become. It was holistic in a way I could never have imagined. This has taken me to a completely new level and exploded my confidence and abilities. What you learn are tools you can use right throughout your life going forward in anything you can think of and even more!"

—SUNIL RAHEJA

"I learned how to use AI to EXPAND my time (not to offload my work). I now have a clear understanding of what AI is, what it is NOT, and how to leverage it to work much smarter."

—JAMES STEPHENSON

"Danny teaches you how to use AI to become better—not cheaper."

—YAAKOV BELCH

"This is hands down the best AI training program that teaches you how to use AI properly. This is not about prompts or hacks, it's about learning how to make AI your thought partner . . . Danny's approach to AI is unique, and you won't find it anywhere else."

—ANDREA NIÑO DE GUZMÁN

"Danny Iny helps you tame your AI Assistant so it's not a sycophantic fan but a sharp tool that challenges you every step of the way. His methods enable you to do more, but better than you ever thought you could. If Danny's methods were widely adopted, we would be less fearful about AI's effects on our cognition, our classrooms, and our democracy."

–SHERYL COE

"Danny has a guided approach to teaching you to think and use AI differently, as a strategic thinking partner. Well, really more as a tool to help you think more deeply and more clearly. It's almost the opposite of how people are using generative AI today. AI is a thinking tool, not (as Danny says) a 'content vending machine.'"

–WENDY CASTLEMAN

"This was far more than a training—it was a true strategic transformation . . . I built a personalized AI operating system that now amplifies how I plan, create, and execute. If you want to stop using AI as just a tool and start using it as a strategic partner, this is the program to do it."

–LINDSAY WELLS

"Danny has a different view on AI. This was not about the low hanging fruit of content, but about a deep professional transformation. I now think differently, have a very skilled advisor for my decisions, and I am going to keep using AI for both my operations and my key business choices."

–ALBERTO BELLE

"It wasn't just about learning how to use AI more strategically (though that in itself was incredible). What struck me most was how it helped me see AI as a way to augment my own thinking—to push past unconscious limitations I didn't even realize were there. The clarity, insights, and new perspective I gained will shape not only how I use AI, but how I grow in my life and nutrition practice moving forward."

–TANYA AUGER

"Danny showed hundreds of people how to use AI to bring out the best of us. Not as a 'magic lamp' that gives us the answers—that simply doesn't work — but as a sparring partner that can bring out, as he says, the genius in us. I believe that every single one who followed his instructions and did the exercises came out transformed. Even those of us who have been using AI intensively for a long time were surprised by the possibilities."

–CARLOS ACCIOLY

"This was next level because it was about developing processes that can be adapted to a range of tasks in personal and professional contexts. The primer work started the process by changing my thinking and I now have the skills to use AI responsibly and effectively. This course showed how we can train our AI by changing our own thinking. Powerful training for next generation working."

–TESSA SAUNDERS

"The key lesson was to use it thoughtfully: not as a replacement, but as a partner that helps raise the bar on quality."

–ANDREA SZABÓ

"I didn't expect this to change the way I relate to AI—but it did. Completely. Danny's AI Clarity Cascade uncovered things I genuinely didn't know about myself, my thinking patterns, and the way I was approaching AI. It dissolved months of frustration in just a few sessions. The prompts worked with a clarity and precision I've never experienced before, and the insights that emerged were surprising, practical, and immediately usable. Most of all, it reshaped my entire sense of what is possible when working with AI and the importance of CONTEXT. Instead of feeling drained or misunderstood, I came out feeling resourced, grounded, and actually excited to collaborate with AI as a strategic partner . . . It is transformative in a way that sneaks up on you—quietly, powerfully, and with lasting effect."

–TANJA BONESS

"Danny trained us to train the AI to train us. Very cool. And very functional."

–JEFF STASIUK

"I came into the training feeling pretty confident about my abilities with AI, but in just the first day it leapfrogged me to a level way beyond what I had achieved on my own. By the time I had finished, I was already delivering a much higher level of value for my clients."

–JEFF COBB

"This is deep learning that merges strategic thinking with practical reality, which leads to the participant's capacity to expand exponentially. The ratio of time spent and benefits delivered is rare and extraordinary."

–CONSTANCE DIERICKX

"This has really been an eye opener into leveraging AI as a thought amplification tool. It's kind of mind-blowing having such deep, personal and highly contextualised conversations with AI, that you might expect from a personal business coach. Never seen anything like it."

–DOMINIC FURNISS

"Whether you're AI curious or AI confident, you will experience a framework that will help you amplify your ability to create and achieve meaningful business and life goals. This isn't a heavily theory-based training—you will get a whole load of work done along the way, along with some invaluable insights about yourself and how you work . . . This gives you a set of resources that can be used over and over to drive strategy and execution in a way that sustains and protects your energy."

–MARY DUGGAN

"I'm a techie. I come from tech. I thought I knew how to use AI well. I read articles, I built automations with AI. This training changed my entire perspective on the potential of AI and how to use it for my business . . . And the best thing—it is done without replacing my brain and uniqueness. It amplifies it!"

–NIR MEGNAZI

"A tour de force training that simplifies one's access to deep use of AI as a strategic partner. Easy to do, clearly taught and huge insights on what to do and what to watch out for. An essential guide for serious, curious creators."

–DAVID SURRENDA

"What this did was compel me to actually dig in honestly into my actual intentions including the ones I had been hiding from myself. Not only that, but I was able to discover potential outcomes I had not even considered before that would help me grow and thrive as a coach and business owner. Truly mind-blowing." –LAURENCE WARD

"In the creative spaces, such as fiction authors and artists, generative AI is a very controversial topic with a lot of people avoiding it to their own detriment. The training provided by Danny Iny on how to utilize ChatGPT specifically in a way that protects the creative process and enhances our own critical thinking is invaluable and in my opinion should be training that anyone using ChatGPT should be required to go through." –TANYA JOY MORGAN

"This wasn't about learning a few clever prompts. It was about learning how to think with AI. I gained deep clarity about myself — how I make decisions, where I hesitate, and what I actually want next. I uncovered what I truly want to build in my business and, more importantly, how to build it . . . Danny doesn't just teach tools. He teaches integration. He teaches discernment. He teaches you how to use AI to amplify your intelligence instead of outsourcing it." –KATHY GOUGHENOUR

"Danny's training is intelligent about the amazing upsides of AI without being a trendy fad-adopter. And it's sober-minded and honest about the very real risks and limitations of AI without being a fear-based resistor of the new technology. Smart insights without the extremes and drama is what I was looking for. Danny delivered!"

–STEVE BECK

"This was a brilliant introduction to the potential of using new tools as a lever rather than a crutch. Myths are explicitly dispelled and the focus is on how to be personally more effective, which I have not seen elsewhere." –DREW NICHOLSON

"This course grounded the use of AI in a human-centered way that eased my fears of becoming obsolete, dumb or having to reinvent myself from scratch all over again. Instead I now realize that my expertise is the fuel that AI just expands on. It can't do anything without me. I also learned how to have an appropriate and safe relationship with AI—the what to do and what not to do that I didn't have to learn the hard way."

–KATHRYN BREWER

"I've been working with AI as a creative collaborator for a while, but this training helped me see how much more intentional and strategic that collaboration can be. The exercises clarified my focus, especially around monetization and audience growth, and reinforced that AI isn't just a tool—it's a thinking partner when used well. I appreciate how grounded and actionable Danny's teaching has been."

–JULIE BELMONT

"The AI strategist training helped to clear up many misconceptions I had formulated, mostly pushed by social media. What many consider to be a fast and easy method to generate content I came to understand as being so much more. In fact, the last thing I would ever use AI for is content creation; it is a thinking partner like no other. 'It is not a time saver but a time expander!' Now I know what Danny meant."

–ADRIAN GALLEY

"When looking at innovations that really make a difference, the approach Danny has developed of using AI as a thinking partner is a Nobel Prize level contribution."

–DR. CHRIS JOHNSTONE

"This helps you move from AI as a producer of poor online content to a strategic partner that helps you do better things at the same time. Little theory, lots of practice—you'll walk away with one finished piece, as long as you do the exercises."

–ROBERTA KINTZI

"If you wish to use AI in its most powerful way—as an ally rather than an information parrot—then take the time to learn Danny Iny's approach. The exercises and lessons enable you to push AI beyond its use as a reference or content source, and they reinforce your role as the boss of feedback tone, relevance to your goals, and alignment with your way of moving through life and business." –LISA MANYOKY

"Danny's AI training blew me away . . . I uncovered next-level clarity, gained fresh perspectives, and found meaningful, energizing paths for action in my business that feel aligned with how I want to live my life . . . Danny was incredibly generous with prompts and real-time guidance that gave us an experiential understanding of what AI can do, ethical ways to engage, and pitfalls to avoid. The immersive, hands-on approach created a depth of learning I never could have reached through self-study. Danny is a gifted teacher. He is clear, calming, insightful, and funny, which made the learning process so much easier." –MYNX INATSUGU

"This training didn't just teach 'AI prompts'—it taught relationship. A relationship with my own standards, my voice, my integrity, and the tool itself. I walked in curious and a little wary of losing my human edge . . . and walked out with guardrails, language, and a repeatable workflow that actually makes my work more mine, not less. The quality was high, the teaching was clear, and the practices were surprisingly uplifting. It sharpened my thinking, upgraded my strategy, and gave me a way to create faster without rushing the sacred parts. This was skill-building with soul." –DEANN BURCH

"Danny Iny's AI training lives in a world of its own. If you think you know all about AI, then bask in the experience of being gobsmacked. I can't recommend it highly enough." –ELLIOT SULLIVAN

Think Bigger and **Build Better**
with Artificial Intelligence

Danny Iny
with Josh Bowen

Published in Canada by Mirasee Press.

Mirasee Press • Mirasee.com

ISBN:
979-8-9916600-5-1 (hardcover)
979-8-9916600-6-8 (paperback)
979-8-9916600-7-5 (ebook)
979-8-9916600-8-2 (audiobook)

Design, layout, and typography by MIST • www.MIST.rocks

Special Sales

Mirasee Press books are available at a special discount for bulk purchases for promotions, premiums, and corporate training programs. Special editions featuring personalized covers, a custom foreword, corporate imprints, and bonus content are also available.

This book is dedicated to the optimistic voice inside each of us, that has the courage to ask "what could go right?"

Contents

Foreword

Up until a few months ago, I thought of AI as something I could probably do without. I didn't need it to create slides or handouts or workshop content. Those are things I could easily do myself. Then I was in a group with Danny Iny, and he began talking about the power of AI. What caught my attention was that he wasn't talking about small tasks, like the kinds of things you might ask an assistant to do. He was talking about something much deeper: AI as a thinking partner—a resource that could push back, test assumptions, and suggest directions you might not have considered. That intrigued me.

I should explain why that mattered. For the past thirty years, my work has focused on why people resist or support new ideas. I've identified three levels of resistance. Level 1 is "I don't get it"—the idea seems confusing, irrelevant, or overblown. Level 2 is "I don't like it"—an emotional reaction, usually rooted in fear. And Level 3 is "I don't like *you*"—meaning there's a trust gap with the person behind it.

When I examined my own reaction to AI before that conversation with Danny, I found all three. At Level 1, it seemed techy and overhyped, and I didn't particularly care to learn more. At Level 2, I assumed it would be too hard and too complicated for me. And at Level 3, the whole thing felt like it should have words like "crypto" or "snake oil" attached to it. The promises were just too big.

Danny's presentation started to change that. Not because he made big promises, but because he described something that actually sounded useful. At the end of his presentation, someone said they wished we had a transcript of that conversation. Danny simply asked AI to clean up the transcript and turn it into a PDF. By the time the session ended, we all had copies of that discussion. It was simple, but I was hooked. I had to see what else AI could do.

Soon after that session, Danny created a program where we explored these ideas more deeply. Each of us added background information to our own AI knowledge base, so the AI had a better understanding of who we were—our work, our models, the questions we were wrestling with. I included material about ideas I've developed over many years, and the challenges I was currently thinking through.

What surprised me was how quickly I was able to start getting real value from AI. I didn't need any special technical expertise. I just needed to engage with AI thoughtfully.

Reading *AI Curious*, I was delighted. The title captures something essential: Our curiosity allows AI to become a powerful resource. Danny explains that if we only give it tasks—"create an agenda," "summarize this article"—AI becomes little more than a glorified vending machine. But, on the other hand, curiosity invites conversation.

If we ask AI to challenge us, it will. And often the results are surprising and get me to reconsider things I had taken for granted. For example, I was designing a short online workshop that I expected would run three hours. AI helped me reorganize the agenda and improve the flow. And that was useful. Then I asked a different question: *What weaknesses do you see in what I've designed? What am I missing?* The response was eye-opening.

It pointed out that I was trying to teach too much, that people wouldn't be able to absorb that amount of material, and that the opening of the workshop needed to engage participants much more quickly. It suggested cutting the introduction, simplifying the focus, and creating more space for people to explore. I ended up distilling the workshop down from three hours to a 90-minute session. And it is far stronger because of my conversations with AI.

What struck me most was how different this felt from receiving criticism from another person. Often when we hear critiques of our work, we have to brace ourselves. But this felt more like a thoughtful colleague building on my ideas and helping me take them further.

That experience helped me appreciate an important insight in this book: the trap of early closure. It's easy to come up with something, ask AI to tidy it up, and call it done. But if we stay curious—if we keep asking questions, exploring alternatives, and testing assumptions—we often discover far better possibilities. As Danny writes, "*The moment conditions shift, the imagination shifts with it.*"

One thing I especially appreciate about this book is that it isn't a collection of tools or step-by-step instructions. Danny suggests that those would likely be outdated before the book ever reached our hands. Instead, Danny offers something more durable: a way of thinking about AI.

He encourages us to move beyond seeing it as a machine that simply follows instructions and instead treats it as a partner in exploration—a resource that can challenge us, expand our thinking, and help us see possibilities we might otherwise miss.

Reflecting on my experience now through the lens of my own framework, I can see what shifted. My Level 3 resistance—trust—was already handled. I trusted Danny long before I trusted AI. That trust is what got me to engage with the book in the first place. The book itself resolved Level 1—I could understand what he was writing about, because he wasn't writing about technology. He was writing about thinking. And Level 2 dissolved through experience—once I saw how AI could help me increase the quality of what I deliver to clients, the fear lost its grip.

If you picked up this book with any of those three levels of resistance in play, I suspect you'll find, as I did, that curiosity is the solvent for all of them. I hope your experience reading it is as powerful as mine has been.

—RICK MAURER

INTRODUCTION

The Case for Curiosity

IN JANUARY 1927, a 44-year-old geologist named J. Harlen Bretz was invited to present his research to the Geological Society of Washington. The room was packed with eminent scientists. Bretz later called them "six challenging elders." Contemporaries described the response to his presentation more vividly. They called it "a lynching."

For five years, Bretz had been studying eastern Washington's bizarre landscape, which included a 400-foot dry waterfall stretching three miles wide and house-sized boulders scattered across the plateau like marbles. He called it the Channeled Scablands, and proposed that this landscape hadn't been carved over millions of years by gradual erosion. It had been created in a matter of days by a catastrophic flood.

The geology establishment was horrified. The prevailing doctrine held that the Earth was shaped only by slow, gradual processes. To argue for sudden catastrophe smacked of the biblical literalism that scientists had spent decades fighting to overcome. After Bretz finished speaking, six geologists rose in succession to demolish his theory. One called it "preposterous and incompetent."

Most of his critics had never actually visited the scablands. They knew the theory couldn't be true, so why bother looking?

Bretz spent the next four decades in scientific exile. He published paper after paper defending his work while the establishment continued to dismiss him. Then, in 1965, a group of geologists finally toured the scablands themselves. They walked the channels. They stood at the edge of Dry Falls, and they saw the evidence with their own eyes.

They sent Bretz a telegram: "We are all catastrophists now."

It would be easy to read this as a story about one stubborn genius who was right while everyone else was wrong. But that misses the point. Bretz's critics weren't stupid. They had coherent reasons for their skepticism. He hadn't yet identified where the water could have come from, and his theory did sound implausible. The problem wasn't their intelligence or their expertise. The problem was that they stopped being curious. They were so certain they already knew the answer that they never bothered to look at the evidence that might have changed their minds.

And the breakthrough didn't come from the new theory or better arguments. It came from humility, from geologists who finally said, "Let's go see for ourselves."

The same thing is happening right now with AI. Most people who've tried these tools and walked away unimpressed aren't wrong about what they experienced. But they're looking at the Scablands from a distance. The gap between disappointing and extraordinary results almost never comes down to the technology. It comes down to whether you were curious enough to stay in the conversation past the point where most people leave. And most people never discover this, because nothing about the default experience suggests there's more to find.

THE LOUDEST CONVERSATION

These days, everyone has an opinion about AI.

Some arrive breathlessly, full of promises about transformation and disruption. Others are dismissive, waving away the whole thing as overblown, a bubble, a passing fad not worth the attention it's demanding.

Scroll through your feed on any given day and you'll encounter both—often within minutes of each other. The enthusiasm is loud, and so is the skepticism. And somehow, neither feels quite right.

In the face of all this noise, uncertainty is a rational response. It's not a sign that you're behind, or that you've failed to grasp something that everyone else has figured out. Anyone who claims to have it all figured out is either dangerously naive or irresponsibly cavalier.

The concerns are real. Many people are appropriately worried about the enormous environmental impact of all the power needed to run these systems. The intellectual property protection concerns are legitimate. The risk of deep fakes, of synthetic media so convincing it becomes impossible to trust what you see and hear, is not theoretical anymore. And then there's the flood of AI-generated content eroding trust across the internet, the irresponsible use of these tools making people who rely on them less capable instead of more, not to mention the very real prospect of massive job disruption.

These aren't hypothetical fears. They're already visible. Anyone who tries to dismiss them with a wave of the hand is missing the seriousness of the moment we're in.

So if you've felt uneasy, skeptical, or simply overwhelmed by how much noise surrounds this topic, that's not a personal failure. It's a contextual one. The conversation has been so loud, so polarized, and so lacking in nuance that feeling confused or cautious is the reasonable response.

What tends to happen in moments like this is that people gravitate toward one of two positions.

The first is full-throttle "AI will change everything" enthusiasm. It will fix problems we didn't know we had. It will unlock productivity, creativity, and capability in ways we can barely imagine. The only appropriate response is to adopt it as quickly and completely as possible, to become an early adopter, master the tools and ride the wave before it passes you by. This posture has a seductive energy to it. It feels proactive and forward-thinking. It's alive to possibility. And there's something to be said for that energy. It comes from a genuine recognition that something significant is happening in the world.

The second position is avoidant skepticism. It comes from the belief that this is all overblown. The hypesters are way over their skis, well beyond what reality can justify. We've seen this before with other technologies, and the promises never quite deliver. Better to wait until the dust settles, until it's clear what's real and what's just noise. There's no need to rush into something that might turn out to be a distraction. This posture has its own appeal too. It feels grounded, measured, immune to hype. And there's something to be said for that too. It comes from a reasonable wariness about being swept up in a trend that might not deliver what it promises.

Both reactions are understandable, and I've felt each of them at different points. But I've come to realize that both of these postures are *reactive.* They collapse complexity too quickly. The enthusiasts assume they know what AI means before they've really sat with it. The skeptics assume they know what it doesn't mean without engaging deeply enough to find out. Both groups have reached their conclusions before they've done the work to understand what they're concluding about. Neither stance creates the conditions for good thinking. And good thinking is exactly what this moment requires.

We need a way to engage that is neither breathless adoption nor reflexive dismissal, taking this seriously without taking it on faith. That's harder than it sounds. It requires a kind of patience that our current environment doesn't reward. But it's worth the effort, because the alternative is locking into a position before you've really understood what you're looking at. And that carries costs that are easy to miss until you're already paying them.

And regardless of how excited or worried any of us might feel about AI, it isn't going anywhere. The genie is out of the bottle. The toothpaste is out of the tube.

I say that not to pressure you or to suggest that you need to catch up. I say it because trying to pretend otherwise doesn't actually preserve the world you're used to. The cognitive environment has already changed. The way work happens is already shifting. And expectations around what's possible are already adjusting. Non-engagement isn't a neutral act. It's a choice that carries its own consequences, even if those consequences are invisible at first. You keep wrestling with problems the hard way—not because you prefer it, but because you don't know there's a different kind of conversation available to you. You make decisions based on the thinking you can do alone, without realizing how much more you could hold in view. From the inside, nothing feels different. Your process feels the same as it always has. That's what makes the cost so easy to miss. None of this feels like falling behind, because the gaps don't announce themselves. They just accumulate. But deciding not to participate doesn't stop the changes from happening. It just means you'll have less say in how those changes affect you.

I've seen people try to opt out of this conversation entirely. Some of them try to shame others into not using AI, as if social pressure could turn back the clock. These reactions remind me of a toddler sticking their fingers in their ears and screaming. It might feel satisfying in the moment, but it doesn't change anything. The world keeps moving regardless of whether you've decided to participate.

I'm not suggesting that you need to panic, or move faster, or master a list of tools before some imaginary deadline passes. Inevitability doesn't call for urgency. It requires orientation. Feeling the need to sprint to keep up isn't the same as acknowledging that conditions have changed. The first is manufactured pressure, and the second is adult realism. This book is interested in the latter.

THE CASE FOR CURIOSITY

Which brings me to the title: *AI Curious*. There are two ways to read that phrase.

The first meaning is straightforward. Being curious about AI. Wanting to understand what it actually is, how it works, what it can and can't do. Wondering why it sometimes feels powerful and other times feels disappointing. Asking honest questions without assuming you already know the answers. This kind of curiosity is essential. It's the starting point for any genuine engagement with something new, and without it, you're limited to whatever conclusions you brought in the door.

But there's a second meaning that goes deeper. Curiosity as a way of showing up in the world. Approaching the subject, and the work you do with it, through a stance of genuine inquiry rather than judgment. And curiosity is, practically speaking, the stance that gets the most out of AI. The people who get disappointing results almost always share the same pattern. They approach it with a fixed idea of what they want, issue a command, and evaluate the output. But the people who get exceptional results do something different. They stay in the conversation. They follow up, push back, offer more context, and let the interaction evolve. They treat it as a thinking partnership rather than a vending machine. And when they do, something changes. They describe moments where the interaction surfaced a connection they wouldn't have made alone, or reframed a problem they'd been stuck on for months. Not because AI knew something they didn't, but because the conversation itself created space for better thinking. They describe it as feeling like their own intelligence just got larger. Because the disposition you bring to AI isn't separate from the results you get. It shapes them directly.

Judgment comes easy. We all have opinions, and when something new arrives on the scene (especially something that seems to threaten how we work or how we understand ourselves) the temptation is to sort it quickly into categories. Good or bad. Revolutionary or overhyped. Threat or opportunity. Judgment gives us the comfort of knowing where we stand. The problem is that it often locks us in before we've had a chance to understand what we're looking at.

Curiosity works differently. It doesn't demand immediate resolution. It's willing to sit with ambiguity long enough to see what's actually there. It asks questions instead of defending positions. This kind of patience doesn't come naturally to most of us. We want answers and clarity. We want to know what this means and how to respond. But true understanding emerges from sustained inquiry, not snap judgments. The willingness to stay in the question longer than feels comfortable separates genuine understanding from confident ignorance.

The people who tend to underestimate things (or overestimate them, for that matter) usually have one thing in common: they weren't curious enough to ask questions before they made up their minds. If they had been curious, they would have discovered that the reality was more nuanced than their initial reaction allowed for. Curiosity protects us from the errors that certainty invites. It keeps us open to information that might complicate our conclusions. And in a moment like this one, where the technology is evolving faster than our understanding of it, that openness is essential.

So the title of this book names more than a topic. It reveals a stance, an invitation to a way of engaging that makes everything else in these pages possible. Because it's a lot easier to learn if we leave our preconceptions at the door.

HOW AI SHAPES US

There's another dimension to this. The way we interact with AI shapes us.

The habits we develop in repeated interactions don't stay contained to those interactions. The patterns of attention, the ways we speak, the expectations we form . . . they all spill over. They become part of how we think and relate more broadly. What you practice becomes who you are, even when you're not practicing.

I'll give you a small example. I say please and thank you to AI. Not because it matters to AI. It doesn't care. It doesn't have the capacity to care. I do it because I think it matters for me. If you get in the habit of talking in a way that you wouldn't want to be talking to people, it's going to start spilling over into how you interact with other people. The way we relate to systems trains habits of mind.

Now, there's actually some research suggesting you get better results from AI if you're rude to it. Maybe that's true today, with the current generation of models. But algorithms and technologies will come and go. What stays constant is you. The person you become through thousands of small interactions doesn't reset when the next model ships. I'd rather protect the human than optimize for the algorithm.

You can see the effects in practice. Barking orders, expecting instant compliance, treating the system as purely instrumental... that all trains impatience and disposability. It cultivates a kind of sharpness that doesn't serve us well when we're dealing with other humans. You might have seen this happen. Someone develops a habit of curt, demanding interaction with AI, and that same curtness starts showing up in their emails, their meetings, their conversations with colleagues. They didn't intend for it to spread, but habits have a way of generalizing. What you practice with the machine becomes part of how you move through the world.

Curious, dialogical interaction trains something different. It emphasizes attention and care, the patience to stay with a question long enough for understanding to emerge. It keeps judgment in the picture rather than outsourcing it. And those habits generalize too, in directions that are far more beneficial.

I'm not suggesting you need to be nice to your computer. But there's something worth protecting in yourself as you learn to think alongside something powerful. That's not a small thing. It might be one of the most important things this book has to offer: a way to use AI that lets you remain fully human while doing so.

WHAT THIS BOOK WILL (AND WON'T) DO

The goal of this book is to help you think more clearly by developing a way of engaging that makes clarity more likely to emerge. You'll learn to approach AI deliberately, with intention, rather than reactively or impulsively. And you'll learn to protect your judgment and your energy, resources that are more precious than any efficiency gain. The way of working you develop will compound over time. The more you practice it, the more it becomes part of how you think.

The point is not to teach you specific tools, catalog apps, walk you through workflows, or help you build agents. And I'm not promising to save you time. The real value of AI isn't as a time saver—it's as a time expander. It doesn't give you more hours. It lets you do better thinking in the hours you already have.

Tools change quickly, and there's a good chance that what is cutting-edge today will be obsolete in six months. If I spent this book teaching you the interface of a particular product, you'd have something that was already outdated by the time it went to press. And even if the tools didn't change, learning to click the right buttons wouldn't give you what you need. Knowing how to operate a tool is different from knowing what the tool is for, when to use it, how to use it well, and when to set it aside. The second kind of knowledge is what matters, and it doesn't come from product tutorials.

The real leverage is in how you think. If you learn to think well with AI, if you develop the stance and the habits that make good thinking possible, then tools become secondary. You'll be able to pick up any tool and use it effectively, because you'll understand what you're trying to accomplish and why. That kind of capability doesn't expire. It transfers and compounds, giving you something much more valuable than proficiency with any particular interface: judgment about what these tools are for and how to use them well.

The ideas in this book were first developed for a training I lead called AI Strategist, where I work with business owners and professionals to figure out what actually makes the difference between disappointing AI interactions and extraordinary ones. One participant described AI going from "a fluff generator" to a "trusted thinking partner." Another said it wasn't thinking for her, but extending her thinking. A third went "from curiosity to real breakthroughs on challenges I'd struggled with for years." That distinction kept coming up, and it's the core of what you'll find in these pages. You could read it just to learn how to use AI so you don't get left behind. But the real goal is something bigger: becoming capable of more than you thought possible.

I'll admit that I didn't set out to develop a philosophy of AI engagement. I wasn't looking for a framework or a methodology. For a long time I was skeptical (maybe even resentful) about the whole thing. There was other work I cared about more deeply, and AI felt like a distraction, an obligation rather than an opportunity.

That all changed because of a particular experience I had with AI. I stumbled into a way of working with it that produced results I hadn't expected, because I was desperate enough to try something different. The circumstances forced my hand, and in being forced, I discovered something I wouldn't have found otherwise.

From that experience, I began to understand why most people's interactions with AI feel disappointing, and why it doesn't have to be that way. The disappointment came from their approach rather than the technology. Most people were treating AI like a search engine with better grammar. They would ask a question, get an answer, then move on. That's not wrong, exactly, but it misses most of what these tools can do. Once I understood that distinction, everything looked different.

So let's start with the story of how I accidentally discovered a different way of working with AI, and what it taught me about thinking.

CHAPTER 1

My Accidental AI Awakening

MY EARLY EXPERIMENTS WITH AI were . . . underwhelming. The outputs weren't exactly *bad*, but they were hollow. I'd ask for a draft and receive text that technically fulfilled the request, but missed everything that would have made it worth reading. The sentences were grammatically correct and the structure made sense, but it all felt empty—like text written by someone who understood the assignment but had no stake in the outcome.

But I could still see how much AI was going to impact our industry, and I didn't want to be caught flat-footed. So I kept at it. I talked to people who knew more than I did and paid attention to the use cases that seemed to get people so excited. I read articles, watched demos, and tried to understand what everyone was so worked up about. And after all of that, I still wasn't super impressed. It felt more gimmicky to me than anything else, like the hype was way ahead of reality. The whole thing felt like a solution in search of a problem.

The truth is that I felt resentful that I had to focus on it at all. There was other work I cared about more deeply—work I'd spent years developing, answering questions that I found genuinely interesting. But everyone wanted to talk about AI! So I kept exploring, trying to find what everyone else seemed to see. I figured that eventually something would click, that I'd stumble onto the use case that would make it all make sense.

All it seemed to be was a glorified content vending machine. For most people, it's a pretty linear process: you give it prompts, and you get back answers. And the results, like what you get from a vending machine, they're not that great—more of a cool party trick than an actual useful business tool. Fast, yes. And impressive at first glance. But if you look more closely, the output is typically thin and generic, the kind of work that would need almost as much fixing as just doing it yourself.

I kept waiting for the moment when it would click, when I'd suddenly understand what all the excitement was about. That moment didn't come until I had an experience that had nothing to do with business.

WHEN THE FRAME BROKE

They say that a healthy man has a hundred wishes, but a sick person has only one. I can relate to that—I'd been dealing with a persistent health issue for a while, the kind that doctors acknowledge is real but couldn't quite figure out. I'd seen specialists and gotten tests. Each new doctor would listen to my history, nod thoughtfully, and offer their best guess. The guesses were smart, but they also weren't leading anywhere. I'd leave each appointment with a new hypothesis to try and a vague sense that we were circling the same territory we'd already covered.

Now, it would be easy to blame the doctors, and sometimes I did. Despite the mythology of doctors as pinnacles of expertise and skill, in real life doctors are just people. I can prove mathematically that 50% of all doctors graduated in the bottom half of their class, and the more experienced they are the further their training was removed from the current best research and practices.

But really, the doctors weren't the problem. They were mostly competent and doing their best. The real problem was structural. A fifteen-minute appointment doesn't leave room for the kind of exploratory thinking that a confusing case requires. Each doctor would listen, form a hypothesis, and move on to the next patient. The process followed all the standard protocols. But for a case like mine, where the answer wasn't obvious and the symptoms didn't fit neatly into a single category, it was insufficient. Because the constraint wasn't expertise—it was time.

I needed a doctor who could spend hours with me instead of minutes—someone who could consider every possibility and keep asking questions until an unexpected idea surfaced. Someone who could hold the whole picture in their head and notice connections that would be invisible in a fifteen-minute snapshot. Someone who recognized that if there were no horses to be found, it was time to start looking for zebras. But that doctor doesn't exist, because the economics of healthcare don't allow for it. Every minute with one patient is a minute taken from another. So there's no villain in this story. It's just the way the system works.

Around that time, I'd been reading Peter Attia's book *Outlive: The Science and Art of Longevity* and rewatching *House MD*, a guilty pleasure from years ago. The combination stuck with me for some reason. Attia's systematic and holistic approach to health, his insistence on looking at the whole picture rather than treating symptoms in isolation, his willingness to dig deeper than conventional medicine typically allows. And the fictional House's relentless willingness to consider every possibility, no matter how unlikely, and his refusal to accept "we've run out of time" as a reason to stop investigating. I was desperate and had nothing to lose, so I figured "hey, let's try something."

THE EXTENDED DIALOGUE

I opened a new chat and gave it an unusual instruction. I told it to act as a combination of Peter Attia and Gregory House. I asked it to assemble a panel of medical subspecialties, more than twenty different perspectives, and then interview me systematically about my symptoms. I wanted questions, not diagnoses. I wanted it to probe and ask follow-ups, rather than conclude.

I'm not sure what I was expecting, but I was completely unprepared for what happened next. The conversation took hours, and then continued the next day and the day after that.

When it had exhausted every question that each of the medical panelists wanted to ask, I told it to give me every diagnosis that could possibly fit the symptoms, not necessarily likely but every possible one, stack ranked in order of likelihood. And then we kept going deeper. This process literally went on for days. Each session would pick up where the last one left off. Each round of questions would surface a thread I hadn't considered, some angle I hadn't explored. AI would ask about symptoms I'd mentioned in passing three days earlier, drawing connections across conversations in ways that felt almost uncanny. It remembered everything. It never got impatient. It never seemed eager to wrap up.

What struck me most was the depth of exploration. With a human doctor, there's always an awareness of time pressure. Even when you see someone with great bedside manner, you edit yourself, skipping details that seem minor because you can tell they need to move on. You compress your messy situation into something that can be processed quickly. Since you know the system, you learn to be efficient, to get to the point and respect their time. But here, there was none of that. AI would ask a question, I'd answer, and it would probe further with follow-up questions I hadn't anticipated. Sometimes the questions seemed tangential, but I'd answer them anyway, and those tangents would often lead somewhere unexpected. It surfaced connections I hadn't made.

AI was patient in a way that no human could afford to be. It wasn't smarter than (some of) the doctors I'd seen, and it didn't (necessarily) have access to knowledge they lacked. The difference was that it could afford to be orders of magnitude more thorough. It could keep asking questions without worrying about the waiting room filling up or the schedule running behind.

I felt that difference, and found myself answering differently than I would have with a human doctor. I shared the details that seemed minor and explored the hunches that might not lead anywhere. I said "I'm not sure, let me think about that" and actually took the time to think, sometimes sitting with a question for several minutes before responding. The conversation meandered in ways that felt inefficient but turned out to be productive. Sometimes the most valuable moments came from the tangents, from following a thread that seemed irrelevant until suddenly it wasn't. I didn't feel like I was wasting anyone's time—I just followed the questions wherever they led.

The conversation kept unfolding, and I started to look forward to the sessions the way you might look forward to a conversation with a thoughtful friend who happened to have encyclopedic medical knowledge and infinite patience. And I recognize that it was strange to look forward to talking with a system that had no idea I existed between sessions. But the experience of the conversation was real, even if the relationship wasn't.

The thinking it enabled was real, and the insights that emerged were real. When you're given permission to think slowly, connections surface that you'd never notice in a rushed exchange. You hear yourself saying things you didn't know you thought, and the situation starts to look different than it did when you began. Possibilities you'd dismissed as impractical suddenly look viable. Connections between things you'd always kept in separate mental categories become obvious. The shape of the thinking itself changes. Dimensions you hadn't considered come into view.

I've since heard many people say this kind of process makes them feel deeply seen. It's this feeling of, "Finally, someone is listening to me." That's what it felt like. AI didn't understand me in any emotional sense (it doesn't have that capacity). But the sustained attention, the willingness to keep going, created a kind of space I hadn't experienced before, a space where I could think without rushing to a conclusion. I wasn't inputting a question and getting answers back from AI. I was thinking alongside it.

When I stepped back to make sense of what had happened, what stood out most was that no human doctor would ever have spent that much time and attention trying to figure out my case. They couldn't, because human attention is zero-sum. Additional context is expensive—it takes time, energy, and often money. And there's always the risk of the other person getting distracted or tired or annoyed. But with AI, the cost of incremental context was zero. You can always add another round of exploration, or dig a little bit deeper. There's no clock running, no waiting room filling up. The constraint I'd been bumping up against for over a decade—the scarcity of sustained attention—had suddenly lifted.

I didn't yet have language for what had happened. I couldn't explain why this experience had worked when so many previous ones hadn't. But I knew something had shifted in how I was using AI, even if I couldn't yet say what.

WHAT CHANGED (AND WHAT DIDN'T)

I want to be clear about what I'm claiming here. AI didn't solve my health issue or make my decisions for me. I still had to evaluate the possibilities it surfaced and decide which threads were worth pursuing with actual doctors. I still had to take responsibility for my own health decisions. The thinking was mine, and the responsibility never left my hands. AI gave me a better map of the terrain, but I still had to choose where to walk.

AI also didn't give me access to knowledge I couldn't have found elsewhere. Everything it surfaced was, in principle, available in medical literature or through consultations with the right specialists. And yes, I did have to double check and validate the details and conclusions. But my ability to engage with the information had shifted. I could hold more of the picture in view at once, and follow the implications without losing the thread. I could come back to earlier ideas days later and pick up where I'd left off. The conversation accumulated rather than resetting every time. I could explore more deeply and spend more time thinking. Not more hours in the day, but more thinking per hour. The quality of my questions improved.

And one day, I realized I could apply the same approach to my work. I brought the same patience to strategic questions I'd been wrestling with for months—and the pattern held. Quick prompts produced quick outputs, and mediocre results. But slowing down and treating AI as a thinking partner changed what was possible. The process itself was valuable, regardless of what came out the other end.

My wife and business partner, Bhoomi, noticed the difference before I could articulate it. Talking to one of our team members, she said that, "When I work with my AI, it feels like a junior VA. When Danny works with his, it feels like a senior strategist."

That was the first time someone had named what I'd been experiencing. The conversations I was having with AI didn't feel like using a tool. They felt like thinking with someone who could hold the full complexity of a problem without losing the thread. We were all using the same technology, and yet it sounded like we were describing two entirely different systems. The difference was in how we were using it, and the context we brought to the conversation. Most people today bring the same mindset to AI that they bring to a search engine or a vending machine. Put in a request, get out a product. That mindset shapes the interaction, and the interaction shapes the results. The tool responds to how you use it.

What I had stumbled onto was completely different, because the limiting factor with how most people use AI isn't the technology itself. It's the interaction. Change the way you engage, and what's possible changes too. I hadn't set out to discover this. I'd been driven by frustration. But somehow, in the process of trying to solve a problem that mattered to me, I'd accidentally found a new way to interact with these tools and a new way of thinking.

Once I understood what was possible, I couldn't go back to using AI the way I had before. The contrast was too stark. Extended engagement produced something valuable, and quick transactions felt hollow by comparison. I kept finding myself returning to the longer, deeper conversations, the ones where I wasn't sure where we'd end up when we started. The difference, every time, was whether I stayed in the conversation long enough for something real to emerge.

But why did this work so well? What was actually happening in those extended conversations that made them so different from the quick prompt-and-response interactions everyone else was having? I didn't have the answer yet. But I knew I needed to find it.

Before You Move On

Most of us approach AI the same way we'd approach a vending machine—put in a request, get back a result, move on. This section at the end of each chapter is a quick recap of the key ideas, so you can check whether the chapter landed before moving to the next one.

This chapter told the story of an accidental discovery—that the difference between disappointing and extraordinary AI interactions has almost nothing to do with the technology. It has to do with how you engage. Key ideas:

- Most people's underwhelming experience with AI reflects the quality of their interaction, not the capability of the tool.
- The health exploration worked because the usual constraints—time pressure, editing yourself, compressing your situation—didn't apply.
- AI wasn't smarter than the doctors. But it could afford to be orders of magnitude more thorough.
- The shift wasn't in what AI knew. It was in how much context it was given and how long the conversation was allowed to develop.
- The distinction between using AI as a vending machine and using it as a thinking partner changes everything about what's possible.

Ready to Experience This For Yourself?

The way of working with AI that you just read about isn't theoretical—it's something you can start practicing right now, even as you continue reading.

I've put together a free mini-course called Install Your Personal AI Operating System—a 4-part protocol that transforms AI from a generic vending machine into a strategic thinking partner. It's built on the same methodology you'll learn throughout this book, and it will give you a hands-on feel for what's possible before you even finish the next chapter.

mrse.co/ai-os

(Or keep reading—the ideas in this book will make the mini-course even more powerful when you're ready for it.)

CHAPTER 2

Context Is the Fuel for Insight

WHY DID IT WORK? That's the question that stuck with me.

I'd had dozens of underwhelming interactions with AI before that one. Same technology, same access. Yet this particular experience had produced understanding that years of specialist appointments hadn't. So what was different? The answer, I eventually realized, had little to do with AI itself. It had to do with the economics of attention.

I mentioned in the last chapter that human attention is zero-sum. But I want to stay with that idea, because I don't think most of us realize how deeply it shapes our behavior.

Think about what happens when you work with a doctor, lawyer, coach, or any other kind of expert. You summarize. You hit the key points. You try to be efficient because you know there's a clock running somewhere, even if you can't see it. Which means that you leave things out. Peripherally relevant information, details that might matter but probably don't. You compress the messy reality of your situation into something that can be processed in the available time. When you work with someone whose time and attention are scarce, you give only what you believe to be the most relevant context.

We internalize this lesson early. Other people's attention is precious and shouldn't be wasted. Get to the point. Leave out the parts you're not sure about. These habits become so automatic we stop noticing them.

But the truth is that there is always more that we could and should say. Complex situations resist compression, and the most valuable discoveries sometimes happen in the tangents—in seemingly irrelevant details that only an expert would recognize as central. But who has time for that? Doctors have other patients waiting, lawyers bill by the minute, and consultants have other clients. So we cut things short. We accept "good enough", because genuine understanding would require more context than anyone has the bandwidth to absorb.

WHEN THE CONSTRAINT LIFTS

Something different happened during this experience. For the first time, I wasn't editing myself. I wasn't compressing. Instead of fifteen-minute bursts, the conversation unfolded over days.

I shared details I would normally have left out and followed tangents that seemed irrelevant (sometimes they were, but sometimes they weren't). I circled back to earlier topics when new connections surfaced. I said things I wasn't sure about and explored hunches I would have dismissed in most conversations because exploring them would have felt like wasting someone's time.

But AI never got impatient. It never glanced at the clock or gave any of the subtle signals that humans give when they're ready to move on. It asked follow-up questions about things I'd mentioned three conversations earlier. It remembered everything, because that's how the technology works. The constraint that shapes every human conversation, the scarcity of sustained attention, simply wasn't there. That was the power of speaking to a system that can hold unlimited context and never gets tired or distracted. You stop filtering. You stop deciding in advance what's worth saying. You just follow the thread.

This was the mechanical explanation I'd been looking for. AI wasn't smarter than my doctors. It didn't have access to better information. But I could give it more context than I'd ever given anyone, and that additional context made better insight possible. And that changed everything, because context is the fuel for insight.

Most people never experience any of this. They interact with AI the way they'd interact with a search engine or a vending machine. Put in a request, get back a result, move on. Which produces exactly what you'd expect: competent but generic output. A response that fulfills the assignment without exceeding it. If you've used AI a few times and walked away unimpressed, this is probably why. You were seeing what AI produces when it has almost nothing to work with.

It's like asking a stranger on the street for advice about a career change. They don't know your history, your values, your constraints, your fears. They give you generic advice that could apply to anyone. "Follow your passion and trust your gut." Not wrong, but useless, because it's not grounded in anything specific to your situation. That's what most AI interactions feel like. It can only work with what you give it.

Most people assume they're seeing AI's ceiling when they're actually seeing its floor. The first output from a cold-start prompt is the absolute worst version of what the technology can do. It's working with almost nothing—of course the results are generic and unimpressive!

But that isn't a reflection of AI. It's a reflection of the scant context that AI had to work with. They were seeing AI operating under severe constraints, asked to produce something meaningful—kind of like judging a consultant based on a five-minute conversation in an elevator.

The real gold didn't come from the first prompt, or even the tenth. It came after hours of sustained conversation, after I'd provided more context than I'd ever provided to anyone about anything. Only then did something valuable emerge. Most people never get there. They give up long before the interaction has a chance to develop. The single most important thing you can do is stay in the conversation.

FROM PROMPTING TO INTERACTIVITY

A prompt is a discrete event: you craft it, submit it, receive a response, and you're done. If the response is good, you got what you came for. If it's not, you try again or give up. Interaction works differently. Each exchange builds on previous ones. Questions lead to answers that lead to better questions. Context accumulates and understanding deepens. The value doesn't live in any single response but in the process of sustained engagement.

I noticed something during those extended conversations: the quality of my own thinking was changing. I was making connections I wouldn't have made otherwise, seeing my situation from angles I hadn't considered. Most people expect outputs from AI: drafts, answers. What I was experiencing was more like augmented cognition. AI was holding context I couldn't hold myself, remembering things I'd said and connecting them to things I said later. It was reflecting my own thinking back to me in ways that revealed patterns I couldn't see from inside.

It's like the difference between trying to solve a jigsaw puzzle in your head versus spreading all the pieces out on a table. When you can see everything at once, connections that were invisible become obvious. You notice two pieces fit together when you see them next to each other. AI was giving me a bigger table so I could see all the pieces.

VOICE AND CONTINUITY

If context is the fuel for insight, certain things follow. The first is that lowering the cost of expression matters. Whatever makes it easier to share more, faster, will increase the quality of what emerges. Anything that reduces friction increases context.

That's why I prefer voice input over typing. And as a writer, that surprised me. I enjoy how writing forces you to organize your thinking before expressing it. But that organization can also be a barrier. Sometimes you want to share something before you've figured out how to say it. Voice preserves ambiguity in a way typing doesn't. You can trail off and circle back. You can share the messy version of an idea rather than waiting until you've cleaned it up.

Typing encourages you to know what you want to say before you say it. Speaking lets you discover what you think in the act of saying it. For complex problems where you don't yet know what the right question is, that discovery process is often where the value lives. Voice doesn't make AI smarter, but it does make a different kind of thinking possible—the kind that happens when you can externalize your thoughts without having to organize them first. Voice input is a kind of permission slip; it gives you license to be messy and think out loud, to share the unfinished version of an idea without first polishing it into something presentable.

The second implication is that continuity matters. If context accumulates, and if accumulated context produces better insight, then resetting that context resets everything. Starting a new conversation means starting from zero. Whatever understanding developed in previous sessions disappears.

And there's a compounding effect. The more AI already knows about you, the more effective each subsequent interaction becomes. This is the snowball that makes extended engagement so powerful. But it only works if context persists. Without memory, every conversation begins cold. The insight from Tuesday doesn't inform Wednesday. You're always starting over.

Imagine working with a consultant who had amnesia. Every meeting, you'd have to re-explain your business, your goals, your constraints, your history. You'd never get past the basics because you'd spend all your time establishing context that should have carried over from the last conversation. That's what using AI without memory feels like. You're perpetually stuck in the first meeting. If context is the fuel, memory is the tank. A tool without memory can only hold what you give it in a single session. That's like running an engine on a thimble of gasoline. You'll never get far enough for the real value to emerge.

I'm not advocating for any particular product or feature set. Tools change rapidly, and whatever I might recommend today could be obsolete by the time you read this. The point is understanding why continuity matters so you can evaluate tools based on whether they support it. The specific implementations will evolve, but the principle won't. But the bottom line is simple: make sure you're using an AI tool that has memory capabilities, and make sure those capabilities are turned on!

THE SHAPE OF A DIFFERENT RELATIONSHIP

Most people interact with AI transactionally. They bring a task, receive an output, leave. There's nothing wrong with this. Sometimes you just need a draft or an answer. Transactional use has its place.

But there's another mode available, one where the interaction itself becomes valuable, where the process of thinking alongside AI changes and upgrades the thinking itself. The accumulation of context over extended conversations produces understanding that no single prompt could generate.

This requires patience and a willingness to share context without knowing whether it will prove relevant. Our entire lives before AI have implicitly trained us to be efficient and get to the point, so the idea of meandering toward understanding feels almost countercultural. And yet that's exactly what produced value in my health exploration experiment. The meandering, the tangents, the willingness to keep going without knowing where I'd end up. It came from the absence of the usual constraints that shape how we communicate with other humans.

The most valuable skill isn't learning how to prompt better. It's learning how to structure an interaction that lets insight emerge. That's about orientation more than technique. It's about approaching AI as a thinking partner rather than an answer machine. The difference sounds subtle, but it changes everything about what becomes possible.

Once I understood the mechanics behind this, I could create the conditions deliberately. And once I could create them deliberately, a different way of working with AI became available. But understanding *how* to work with AI is different from deciding whether you *should*. There are good reasons to remain skeptical. The concerns are real: environmental costs, job disruption, eroded trust, the cognitive atrophy that comes from outsourcing too much. These aren't irrational fears. That's what we'll explore next.

Before You Move On

The previous chapter described what happened. This one explained why. The mechanical insight is simple: context is the fuel for insight, and most people drastically underfeed their AI interactions because a lifetime of human conversation has trained them to compress. Key ideas:

- Human attention is zero-sum. We've learned to summarize, get to the point, and leave out anything we're not sure about. Those habits carry over to AI, where they're unnecessary and counterproductive.
- Most people are seeing AI's floor, not its ceiling. A cold-start prompt is the worst version of what the technology can do.

continued next page

- The shift from prompting to interaction is the shift from discrete events to accumulated context. The value lives in the process, not in any single response.
- Voice input lowers the cost of expression and lets you share the messy version of your thinking before you've organized it.
- Continuity matters. If context is the fuel, memory is the tank. Without it, every conversation starts from zero.

CHAPTER 3

The Shape of Our AI Future

BEFORE WE GO ANY FURTHER into the immense power and promise of AI, we should pause to consider the dangers. Because if you feel uneasy about where this is all heading, you're not being paranoid.

Consider the matter of environmental impact. The energy required to run large-scale AI systems is staggering, creating massive drains on power grids and driving up the cost of electricity for millions of people. Immense amounts of water are used to cool those data centers and power plants, threatening local water supplies, sanitation, and ecosystems.

Then there's the question of who owns what. These models were trained using the scraped work of creators who never consented to its use, and their outputs can blur the line between synthesis and theft. When a system can produce text or images in the style of an artist whose work it absorbed, attribution becomes hard to track and ownership difficult to defend.

The erosion of trust runs deeper still. Deep fake technologies make fabricated video indistinguishable from reality. Voice cloning makes phone calls from loved ones impossible to verify. AI-generated images are becoming difficult and sometimes impossible to distinguish from the real thing. Not to mention the abundance of low-effort AI-generated content flooding every platform and feed, drowning signal in noise. The capacity to manufacture convincing lies at scale is no longer hypothetical.

And perhaps most troubling of all is the risk to thinking itself—the quiet erosion that happens when people let machines do the cognitive work that used to keep their judgment sharp. All of this is unfolding against a backdrop of massive job disruption, with roles changing or disappearing faster than social systems can adapt. Skills that took decades to develop can become less valuable overnight.

Anyone who tries to dismiss or minimize these concerns is either dangerously naive or irresponsibly cavalier. If you look at the world clearly and feel pessimistic about what AI might mean for it, that is not fear—it's pattern recognition. I'm not going to tell you these concerns are overblown. They are real. But wishing things were different will not change what is already happening.

THE GENIE IS OUT OF THE BOTTLE

AI capabilities are no longer rare or centralized. They're not locked in research labs waiting for permission to spread. They're embedded in everyday tools, accessible through consumer products, and woven into workflows across industries. The most capable models in the world are available to anyone with an internet connection and a few dollars a month. Meanwhile, regulation is lagging. Norms are still forming, and ethics committees are debating principles while deployment accelerates. The gap between what is possible and what is governed keeps widening.

Choose whatever metaphor you prefer. The genie is out of the bottle. The toothpaste is out of the tube. Pandora's box is open. There is no realistic path to putting it back. Whether we like it or not, the environment has changed. And the reactions of some people and sectors to either ignore AI or try to shame others into not using it don't actually slow anything down—they just mean those people have less influence on where things go.

As we hurtle toward a post-AI world, there are two common stances that are being adopted. They both seem reasonable, but neither leads anywhere good.

The first dead end is to opt out. This is the head-in-the-sand approach, and it sounds like, "I can't deal with this right now, I have important work to do," or "this doesn't really apply to me." And I understand the appeal. There's wisdom in not chasing every new fad or technology. Like I shared earlier, I felt resentful myself when I couldn't avoid the AI hype. But the problem with opting out is that it doesn't actually preserve anything. The world you wish you could hold onto is already changing. Expectations are shifting regardless of whether you participate. Competitors are recalibrating, and customers are updating their perceptions and expectations of what is possible and what is normal. Ignoring AI doesn't preserve the world as you like it, it just means you will be the last to notice it has already changed into something else.

The second false exit is using AI badly. This is subtler, and in some ways more dangerous. Using AI badly looks like convenience-first delegation. Asking the machine to think, so you don't have to. Accepting the first output without scrutiny because reviewing it feels like extra work. And the seduction is real. Using AI this way feels productive; you move faster, the backlog clears, and outputs multiply at a dizzying pace. From the outside, and often from the inside, it looks like efficiency.

But every time you accept an answer without interrogating it, you skip the cognitive step where understanding forms. When you let the system summarize before you have synthesized, you lose the chance to notice what did not fit the frame. Delegating judgment rather than using AI to inform it leaves the judgment muscle unused. And cognitive muscles weaken when they are bypassed. Confidence replaces comprehension, and judgment is slowly eroded by a thousand small moments of not bothering to think it through yourself.

Ignoring AI does not preserve intelligence. But using AI badly actively degrades it. And if this is what you've been doing, it isn't a personal failure. Systems shape cognition; if the tools around you reward shallow engagement, shallow engagement becomes the path of least resistance. It's a structural problem, but that doesn't make the consequences any less real. Over time you are becoming less capable, not more.

WHY I'M OPTIMISTIC

If we don't really have a choice about whether to engage (because that is already determined), then we should carefully choose the stance we take while engaging. The change is happening with or without us, but we do get to choose how we relate to it. The sense that you must figure everything out immediately or be left behind forever—that's not mandatory. It's just one possible response among several.

I'm actually optimistic about the future, and it's not (just) because I have a cheerful disposition. I'm not wearing rose-colored glasses, as I hope the previous section has shown. The reason why I'm optimistic is that we've seen this all before.

The longer view of history shows us that every major technological and social transition has produced chaos alongside opportunity. The printing press spread literacy and also spread misinformation, destabilized power structures, and fueled religious wars that killed millions. Industrialization increased prosperity over time, but it also displaced workers, created new forms of exploitation, and concentrated power in ways that took generations to address. The internet connected people across the globe while simultaneously fragmenting attention, weaponizing information, and creating vulnerabilities we are still learning to navigate. Each of these transitions created moral panics, job displacement, erosion of trust, and new forms of harm. Each required intentional human effort to steer toward constructive outcomes. Progress has never been automatic. It has always been argued for, built toward, and defended.

And yet, when all was said and done, every one of these innovations ended up being positive sum. The pie kept growing. The disruptions were real, but the capabilities that emerged also created possibilities that hadn't existed before. The printing press eventually produced widespread literacy despite the wars it fueled. Industrialization eventually produced prosperity despite the exploitation it enabled. The internet eventually connected billions despite the fractures it created. The pattern holds even when the path is ugly.

(And to those who argue that AI is different than anything that came before it, I agree—but let's also recognize that the exact same thing was said about everything that came before.)

And across centuries, cultures, and systems, there has consistently been a preponderance of people who care about reducing suffering, who want to build fairer systems, who invest energy into solving problems rather than exploiting chaos. That group is not always loud, nor is it always in power. But it has been persistent, and—in the long arc of history—it has always won.

Short-term fluctuations notwithstanding, throughout the entirety of human history, things have only gotten better. Not because some invisible hand guarantees progress, but because enough people keep showing up to push in better directions. The future looks bright not because everything will automatically be fine, but because history suggests that when the stakes are high enough, enough people step forward to shape what comes next. What we are seeing now (the confusion and misuse and backlash and erosion of trust) is consistent with early-stage transitions. There is a lot of turbulence. But I believe it's a sign of change, not a verdict on the destination.

THE BUDDING OF SECOND-ORDER EFFECTS

One of the most important lessons from history is that the biggest effects of any major technology are rarely the obvious ones. The first-order effects (what the technology does directly) tend to be easier to predict. The second—and third-order effects (what the technology makes possible) are where the real transformation happens.

Consider what happened in 1830 when Edward Budding, an English engineer, patented a machine for cutting grass. Before his invention, a manicured lawn was a luxury. You needed workers with scythes, and the results were uneven at best. Budding had been studying the rotating blades used in textile mills to trim cloth, and he realized the same principle could work on grass. His mechanical mower was not elegant, but it worked.

The first-order effect was predictable. Maintaining grass became cheaper and less labor-intensive. Estate owners saved money. Some groundskeepers found themselves out of work. If you had asked anyone at the time what this invention meant, they would have talked about lawn care. Nobody was thinking about what would come next.

But uniform, well-maintained playing surfaces had been nearly impossible to create at scale. Budding's invention changed that. Within decades, large flat fields became economically feasible, and lawn sports that had been informal pastimes began to organize. Football clubs formed across England. Cricket grounds expanded to accommodate growing crowds. The infrastructure for spectator athletics emerged not from any deliberate plan but from the downstream effects of a machine designed to cut grass. There is no straight line from Budding's patent to the NFL. That connection only becomes visible in hindsight, after the constraint has been removed and the consequences have had time to compound.

Economists call this the Budding Effect—named after Edward

Budding, but also fitting with the evocative image of buds that have yet to bloom and reveal their shapes and colors. When a technology removes a barrier, the most significant changes often appear in places no one thought to even look, solving problems no one knew they would have. Even though none of us can predict the future, there's no reason to believe the same pattern won't apply to AI. The direct capability (generating text or images or code) is impressive but not transformative by itself. What becomes possible when that capability removes friction elsewhere is something else entirely.

Consider what could happen when the pressures AI creates collide with the responses those pressures provoke. Take energy. Right now, AI's enormous power demands are a legitimate environmental concern, but those demands are also driving billions of dollars into clean energy infrastructure, nuclear power research, and grid modernization at a pace that climate policy alone never achieved. At the same time, there's enormous competitive pressure to make the models themselves more efficient, to do more computation with less energy. So there's reason to believe those two forces could converge in ways nobody planned for, where the infrastructure built to power AI ends up producing abundant, affordable clean energy that benefits everyone, while the models that triggered the buildout need far less of it than anyone projected. The problem doesn't just get solved. The solution creates something better than what existed before the problem arose.

Or consider what happens when expert-level synthesis becomes available at near-zero marginal cost. A doctor in rural Malawi gains diagnostic depth comparable to a major research hospital. A researcher in Nairobi surveys published knowledge with the same fluency as someone at MIT. Legal aid, agricultural science, engineering, education—in every field where expertise has been concentrated in wealthy institutions, the same pattern plays out. The knowledge was always there. What was missing was a way to make it available where the experts weren't. And when you put more diverse minds on hard problems, the breakthroughs that follow tend to be the ones nobody predicted.

None of this is guaranteed, and I want to be honest about that. These are possibilities, not predictions. But the Budding pattern suggests that the most important consequences of a new technology are precisely the ones you can't plan for, the ones that emerge when a constraint disappears and people start doing things that simply weren't feasible before. AI's biggest impacts will not be the obvious ones. They will be what becomes possible once the constraints of human attention, expertise, and institutional access are loosened in places where they've been tightest.

EARLY MOMENTS REWARD THOUGHTFUL HUMANS

I should note that the strategies I teach in the rest of this book are valid regardless of whether you buy into my optimism or not. But I hope what I've shared in this chapter lays at least some of your fears to rest. Not so you don't advocate and work for responsible change, ethical technology implementations, or regulatory guardrails—I hope you do all those things, because that work is vitally important. But I hope you don't let that important work hold you back, because there is another historical pattern worth noting. In the early stages of any major technological transition, before the technology becomes truly universal, there is a window where thoughtful engagement outperforms raw adoption. The people who use new capabilities with care and judgment create disproportionate value compared to those who simply adopt without discrimination.

You can see this playing out already. Two people use the same AI tool. One asks a quick question, gets a generic answer, and moves on unimpressed. The other spends twenty minutes providing context, asks follow-up questions, pushes back on shallow responses, and walks away with something genuinely useful. Same technology, same access, completely different outcomes. The difference is not in the tool, but in the quality of engagement.

To quote William Gibson, "The future is here. It's just not evenly distributed." My hope is that eventually the approach I'm laying out in this book will be widespread, mainstream, and perhaps even built into the way AI is designed to operate. But, at least as of this writing, these strategies for thinking *with* AI rather than outsourcing *to* AI remain rare—even among those who pay lip-service to the idea. That means there is still a significant advantage available to those who approach this thoughtfully. And this is where my optimism becomes personal and grounded. Not about AI in the abstract, but about what becomes possible for someone who engages with curiosity and care. The person who learns to think with AI positions themselves differently than the person who treats it as just another piece of software.

There's a particular feeling I've come to associate with this stance. Not excitement, exactly. Not the breathless enthusiasm of someone who thinks they've found the next big thing. And certainly not the grim determination of someone bracing for impact. It's closer to what you feel when you finally stop fighting a current, and start swimming with it. The water is still moving. You're still in it. But you're no longer exhausting yourself pushing against something that was never going to stop.

The concerns I outlined at the start of this chapter haven't gone away. They're still real, and we must take them seriously. Environmental strain, intellectual property erosion, synthetic media, degraded trust, cognitive atrophy, job displacement, and lots more. None of these disappear when you decide to engage. They're part of the landscape now, and they'll stay part of it for a long time. But engaging thoughtfully with AI doesn't make you complicit in those harms, any more than disengaging protects you from them (it doesn't). Participation is already happening, whether you've consciously chosen it or not. What matters now is what kind of participant you choose to be.

The people I've watched engage well with AI share something in common. They aren't the ones rushing to adopt every new tool or optimize every workflow. They aren't chasing productivity gains or trying to replace their own thinking with something faster. What they share is a kind of deliberate patience, a willingness to stay curious longer than feels efficient, to let understanding develop before reaching for conclusions.

That's the stance this book is built around. Not enthusiasm, and not fear. Rather, it's a calm confidence, grounded in the recognition that the most important skill isn't technical mastery or "prompt engineering" (which you'll see isn't nearly as big a deal as most people think). It's learning how to structure your thinking so that insight has room to emerge, and to work with AI in a way that develops judgment rather than bypassing it. Which brings us to the question of how, exactly, that works.

Before You Move On

This chapter made the case that the concerns about AI are real—environmental strain, intellectual property erosion, synthetic media, job disruption—and that dismissing them is naive. But it also argued that the historical pattern of major technological transitions gives us reason for grounded optimism. Key ideas:

- Both opting out and using AI badly are dead ends. The first doesn't preserve anything. The second actively degrades your thinking.
- Every major technological transition has produced chaos alongside opportunity, and every one has ended up being positive sum—not automatically, but because enough people showed up to push in better directions.
- The Budding Effect tells us that the most important consequences of a new technology are the ones nobody predicts—the second- and third-order effects that emerge when a constraint disappears.
- In the early stages of any major transition, thoughtful engagement outperforms raw adoption. There is still a significant advantage available to those who approach AI with curiosity and care.

CHAPTER 4

The Clarity Cascade

IMAGINE YOU'RE TRYING TO SOLVE an important business problem, but you can't seem to figure it out. For example, let's say that your flagship offer isn't converting—but the landing page looks good, the copy is solid, and the price point is competitive. So you sit down with a strategist, someone you trust, and lay out the situation.

She asks, "When you talk to people who almost bought but didn't, what do they say?"

You shrug. Most of them say the offer sounds great—they're just not sure it's for them right now. You've been reading that as a timing issue. Maybe they're busy. Maybe the economy feels uncertain. Maybe they'll circle back later. She doesn't contradict you. She just asks another question: "What if it isn't really about timing? What could be underneath that excuse?"

You pause. You hadn't considered that. If it's not timing, then what is it? "They might not see themselves in it yet," you say slowly. "They know something isn't working, but they haven't named the problem."

That shifts something. You start describing your audience differently. Not as people ready for a solution, but as people still trying to understand what's wrong. The conversation moves from conversion rates to awareness levels. From copy tweaks to audience maturity.

A few minutes later, another question lands. "If they haven't identified the problem yet, what are you asking them to do?"

Now the pattern becomes clearer. You're asking people to commit to a solution before they've fully understood the problem. The issue isn't price. It isn't even positioning in the narrow sense. It's sequence. You've structured the journey in the wrong order.

Nothing new entered the room during that conversation. The data didn't change. The testimonials didn't change. The landing page didn't change. What changed was how you were interpreting information you already had. Each question didn't deliver an answer—it shifted the frame just enough to make the next question possible.

You walked in thinking you had a conversion problem. You walked out understanding you had a sequencing problem. The facts hadn't changed. The frame had.

I've come to call this process—the progressive reorganization of a problem as context accumulates—the Clarity Cascade.

We've all experienced this. A long conversation with a mentor where the real problem didn't surface until forty minutes in, or a negotiation that seemed to be about pricing until you realized it was really about trust. Clarity wasn't achieved just by adding new information—it only came when the process of exploration led you to re-see what was already there. An assumption you didn't know you were making got surfaced and challenged, and once that shift happened, everything that came before looked different.

These moments feel like happy accidents—a flash of insight or a lucky question. But they aren't. Rather, they are the predictable result of context accumulating long enough for the frame to shift. A shift in interpretation reveals what was already present but unnoticed, and once that shift happens, different questions become possible. Each loop can reveal what was invisible before, the way increasing the resolution on a photograph reveals not just sharper features but a second person in the shadow, a sign in the window that changes the context of the whole scene.

Most people assume clarity comes from adding more information. Gather enough facts, and the answer appears. But that isn't what happened in that conversation. The information didn't increase. It reorganized. What changed wasn't the data, but the frame through which it was being interpreted. And when the frame shifts, the questions shift with it. Additional context doesn't just sharpen what you already see. It can change what you're looking at entirely. That's what makes the process recursive rather than linear.

WHY CLARITY SHIFTS ARE SO RARE

So why don't most people experience this more often?

In part it's because, as we've explored previously, human attention is zero-sum. Every conversation has a clock running somewhere. So we summarize, get to the point, leave out the details that might matter but probably don't. These habits are so ingrained we stop noticing them.

Another misconception is that clarity comes from thinking harder or faster. But clarity doesn't come from speed. It rewards duration. Insight rarely appears in the first pass. It emerges when you stay with a problem long enough for the frame to shift. What matters isn't how quickly you think. It's how long you remain in the loop.

This is exactly what happened with my health situation. For years, every doctor got the same thing. A summary that could be processed in the time available. The key symptoms, the highlights of my history, and my current best guess as to what was going on. It was efficient and organized, but precisely the wrong approach for a case that needed more room to breathe. The details I left out, the tangents I didn't follow, and the hunches I dismissed as probably irrelevant—that was the material that would have changed the whole picture. Once I let the conversation expand, the understanding that emerged wasn't just deeper, it was *different*.

That's the power of AI—the first conversational partner where these constraints don't have to apply. You can always provide one more piece of information, explore one more tangent, or revisit something from three conversations ago. There's no clock running, and no pressure to wrap up. Because AI holds everything in the conversation, context accumulates rather than resetting, with details resurfacing when they become relevant even if you mentioned them days earlier. In most human conversations, the process stops as soon as something sounds coherent. With AI, there's no natural stopping point. The cascade can continue until the frame truly stabilizes.

Now, to keep the cascade productive, you need patience. But you also need the right questions. Because questions pull context into the process and keep it moving.

When someone asks you a question you weren't expecting, it also redirects attention toward something that was available but unexamined. The answer surfaces context that wasn't yet in play, and that new context changes what the next question can be. This is why questions matter. Not because they are clever, but because they redirect attention. A well-placed question doesn't just gather information. It shifts what becomes visible. And once attention shifts, the frame can reorganize.

Premature answers do the opposite. When you accept a conclusion before the territory has been fully explored, you collapse the space that further questions would have opened. The context that might have arrived stops arriving, because there's no longer a reason for it to. And you'll never know what you missed, because the questions that would have revealed it never got asked.

Every time you accept an answer too quickly, the cascade collapses. Every time you stay in the conversation a little longer, the frame has a chance to shift. This is why the discipline of asking before answering matters. Not as a personality trait, but as a stance toward how you engage with complex problems.

THE TRAP OF EARLY CLOSURE

Now, if the Clarity Cascade produces such valuable insight, why is it so easy to miss?

Imagine you ask about a business decision, and AI gives you a competent-sounding analysis with three considerations and a recommendation. The prose is clean, the structure is logical, the conclusion seems reasonable. So you move on, never even realizing how much you are missing—because what's missing is invisible. The nuances of your specific situation that would have changed everything. The assumptions embedded in your question that went unexamined. The adjacent considerations that would have surfaced if you'd stayed longer. The answer looks complete, but it just isn't—not even close.

Most people don't end their conversations too early out of laziness, but because the early responses seem sufficient. AI is good at producing plausible output quickly, and that capability is precisely what creates the trap. When the first answer sounds good, there's no obvious reason to continue. So you don't. And the deeper understanding never develops.

Most bad decisions aren't wrong in the sense of contradicting available evidence. Rather, they're under-contextualized, conclusions drawn from a picture that felt complete but wasn't. The information that would have changed your entire focus never entered the conversation because it ended too soon. And the discipline of staying in the process longer than feels necessary is how you protect against this.

Allowing the cascade to run longer changes the game, but it doesn't run on autopilot.

Each loop requires human judgment. AI reflects back a synthesis of what you've shared, and you're evaluating whether that synthesis captures what you meant or distorts it, whether the framing helps or misleads. You decide which questions are worth pursuing, which tangents are productive versus dead ends, and what deserves attention and what can be set aside. As additional context arrives, you evaluate its relevance. Does this change the picture? How does it connect to what came before?

Most people see working with technology as automation—an assembly line where the whole point is removing judgment from the process. The Clarity Cascade is the opposite. It demands more judgment, not less. Every loop requires you to evaluate, decide, and recalibrate, and over time you get better at recognizing what matters and distinguishing genuine insight from plausible-sounding noise. Far from bypassing human thinking, this process exercises it. The support AI provides comes from holding context and sustaining the process, but the judgment remains yours.

But all of this raises a question we haven't examined yet. Are we actually bringing the right material to these conversations, or are we solving the wrong problems entirely?

Before You Move On

This chapter named the core mechanism of the book. The Clarity Cascade is the progressive reorganization of a problem as context accumulates—the process by which each round of exploration shifts the frame just enough to make the next question possible. Key ideas:

- Clarity doesn't come from adding more information. It comes from the frame shifting—a change in how you interpret what you already know.
- These shifts are rare in normal life because human attention is zero-sum and conversations have clocks running. AI removes that constraint.
- The trap of early closure is that AI produces plausible output quickly, which makes it easy to stop before the real insight emerges. Stay in the conversation.
- Premature answers collapse the cascade. Staying with the question keeps it open.
- The Clarity Cascade demands more human judgment, not less. AI holds context and sustains the process, but the evaluation is yours.

CHAPTER 5

How AI Expands Imagination

THE MOVIE *LIMITLESS* stars Bradley Cooper as Eddie Morra, a man given access to a secret performance-enhancing drug called NZT-48. He takes a pill and suddenly everything clicks. He thinks faster, sees further, and easily solves problems that had been opaque for years. "I wasn't high," he says. "I wasn't wired. Just clear. I knew what I needed to do and how to do it."

Of course it's a movie, and the idea of a magic pill that can do all this is a fantasy. But the feeling it describes isn't fantasy at all. Most of us have been there: a presentation where everything came together, a negotiation where you could see three moves ahead, a creative session where ideas arrived faster than you could capture them. Moments of unusual clarity where problems that seemed intractable suddenly felt solvable. These moments are rare, but they aren't imaginary. When they happen, we tend to attribute them to circumstance, to a good night's sleep or the right mood or being "in flow." We treat them as lucky breaks rather than glimpses of something that belongs to us.

But here's what stuck with me about that movie. The pill doesn't *add* anything new. It just *unlocks* what was already there. So why couldn't he access it before?

THE INVISIBLE CONSTRAINTS OF THE THINKABLE

There's a parable about a couple of young fish swimming along when an older fish passes by and asks, "How's the water?" The young fish shrug and keep swimming. Later, one fish turns to the other and asks, "What the hell is water?" The point of the parable is that it takes wisdom to see what we're "swimming" in. Most of the time, it's invisible.

So, what are we all swimming in? It's our sense of what's feasible—formed by experience, resources, and what we've seen work and fail. It quietly determines which ideas become thinkable and which ones never form at all. From the inside, it doesn't feel like a limitation. It feels like clear thinking. Only ideas that fall within what feels possible become candidates for serious consideration. The rest don't get rejected. They never arise.

When we try to explain why we don't think bigger, we reach for visible answers—fear, self-doubt, lack of ambition. But these are the wrong kind of explanation. You feel fear. You wrestle with self-doubt. These are things you can see and work on. And yet many of the most driven, confident people I know still operate well below their capacity. They don't feel afraid or stuck. They feel fine—and that's the tell. Whatever is constraining them isn't visible. It's structural. It's the water.

Think of it this way. Before the Wright Brothers, nobody could even conceive of the idea of a quick trip across the Atlantic. It wasn't that people would think about it and reject it as impractical. The idea simply didn't exist, because feasibility defines the boundary of imagination—before you even start imagining!

Now, this isn't a flaw. These constraints evolved to protect you, to save you from a life of pointless rabbit holes that lead only to frustration and failure. The problem is that what feels impossible is often just impractical under current conditions. But "impractical" and "impossible" are not the same thing. When we collapse the two, whole categories of possibility disappear from view. Your system quietly treats today's constraints as permanent truths. It doesn't ask whether the conditions themselves might change.

WHEN THE CONSTRAINTS FALL AWAY

There is a thought experiment that I take students through when I teach this concept. I invite them to think about someone special to them: a parent, a partner, a close friend, a mentor. Someone who they cherish deeply. Then I invite them to imagine creating a special experience with this person. Something meaningful, something that would matter to both of them. What would that look like?

Take a moment to picture it for yourself. Who is your person? What would you do? Where would you go? What would make it special?

If you're like most people, you've already started making assumptions. You're probably thinking about restaurants you know, activities that fit into an evening, things that are achievable given your life as it is. You're operating within a set of assumptions so natural you may not have noticed you were making them. Now imagine a different scenario. You have a few months to plan and a few thousand dollars to spend. What does the special experience look like now? The dinner might become a weekend away. Travel enters the picture. Something carefully designed instead of convenient.

Now imagine the constraints stretch further. You have years to plan and resources that feel meaningfully abundant. What changes? Maybe it's hiking Kilimanjaro together—something that would require training, time, and coordination. Maybe it's tracing a family lineage across continents. Maybe it's creating something together—recording a song with an artist they admire, collaborating with someone who previously felt inaccessible. As the constraints loosen, the ideas don't just scale. They shift categories.

Notice what's happening. You're not becoming more creative. You're operating inside a different boundary of what feels possible.

When I've run this exercise with groups, the pattern is consistent. The moment the conditions shift, the imagination shifts with them. Not gradually, but categorically. What's striking isn't that the ideas get bigger. It's that the original ideas were quietly shaped by constraints no one had named. Time. Cost. Energy. Access. Legitimacy. No one said, "I can't afford that" or "That's unrealistic." They didn't need to. Those boundaries were already built into what felt thinkable.

We like to talk about "thinking outside the box." But the real challenge isn't stepping outside the box. It's noticing that the box exists at all. When constraints are invisible, they don't feel like limits. They feel like common sense.

The constraints weren't in the instructions, they were internalized long before the exercise began. And when the conditions shifted—even hypothetically—something subtle but decisive happened. The version of you that operates within a different feasibility landscape stepped forward. No new skills were required, no external facts were introduced, and nothing about who you are changed. The boundary moved—and with it, your imagination.

What shifted when the constraints were lifted? The easy answer is that people got permission to dream bigger. But that misses the mechanism. Nobody decided to think bigger. What changed was what felt *possible*. At the beginning of the exercise, your imagination operated inside your current resources and commitments. By the final round, when those constraints had shifted, new possibilities appeared—not because you summoned them, but because there was now room for them to exist.

If the barrier were courage, the solution would be motivational: believe in yourself, push through fear, cultivate a growth mindset. Those things have their place, but they don't address what's actually happening here. The constraint isn't emotional. It's structural. When conditions change, so does imagination. Your capacity for bigger thinking was always there. The ideas just need room to form.

THE ARCHITECTURE OF FEASIBILITY

Normally, the boundary of what feels possible only shifts under unusual circumstances. A crisis can do it. When the stakes are high enough, the cost of inaction overwhelms the usual constraints, and you may act with a level of clarity and boldness that usually isn't accessible. But crisis isn't sustainable. It's narrow and exhausting.

Fantasy can shift the boundary too. When you're daydreaming, feasibility temporarily extends in all directions. But the expansion collapses the moment you try to move from imagining to doing.

Sometimes external scaffolding produces the shift. A demanding investor, an ambitious business partner, or a coach who helps you reframe what counts as a constraint. When these work, it's not because they make you braver. It's because they alter the conditions that define what feels possible.

Those conditions tend to cluster around three structural dimensions:

The first is strategy. Strategy is seeing a path from here to there. If you can't see a plausible route to the destination, the destination doesn't meaningfully exist on your map.

The second is execution. Execution is walking that path. Even if you can imagine where you want to go, you must believe in your ability to actually take the necessary steps—the coordination, logistics, and effort required to make it real.

And the third is energy. Energy is sustaining the walk long enough to arrive. You might see the path, and even be able to take the first few steps. But if the cost of the journey is more than you feel you can carry—the mental load, the uncertainty, the sustained effort—the goal is unlikely to be achieved.

These three dimensions together define what becomes thinkable, and therefore what becomes achievable. When strategy feels unclear, the idea never fully forms. When execution feels impossible, the idea has nowhere to land. And when the energy cost feels unsustainable, the idea will never be adopted.

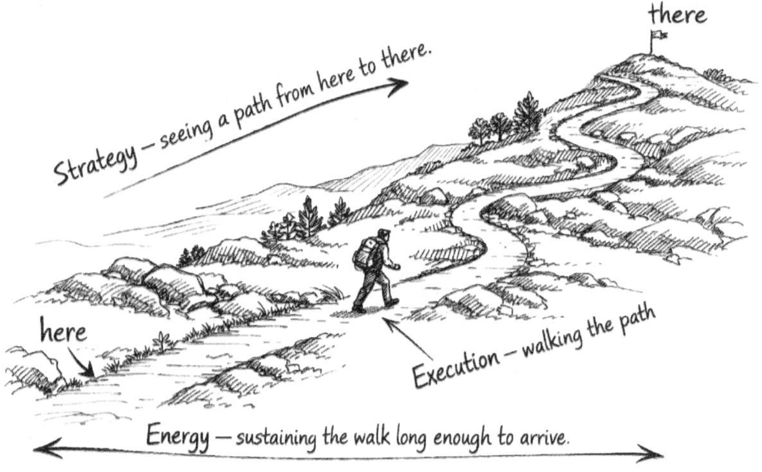

Imagine, for example, starting a company in a new industry. If you can't see how you would begin—who to call, what to build, how to test—the idea doesn't feel real. That's strategy. If you can see the outline but can't imagine assembling the team, building the product, or raising the money—that's execution. And even if you can see and begin, if you can't imagine sustaining the uncertainty and pressure for years, the idea never fully takes shape. That's energy.

This isn't just about business. The same three dimensions shape what feels possible in every domain of life. Planning a dinner party. Training for a marathon. Negotiating your salary. Buying your first home. Changing careers. The question is always the same: Can I see a path? Can I imagine taking the steps? Can I sustain the cost long enough to get the result? When any one of those feels out of reach, the idea rarely gets rejected. It simply doesn't surface as a serious option.

Together these three dimensions define the frontier of what feels possible. You don't experience them as explicit or distinct questions. Rather, you experience them as the range of ideas that occur to you at all. The boundary is already set before any particular idea forms. Which means the ideas that show up—the ones that feel like all the ideas there are—have already been shaped by a landscape you can't see.

Sometimes the boundary shifts on its own—in crisis, in fantasy, or under external pressure. But those shifts are unstable, and hard to summon on demand. Which is why most of us stay inside the same feasibility landscape for years at a time.

AI AS STRUCTURAL LEVERAGE

What if the conditions themselves could change—without crisis, without fantasy, and without external pressure? That's where AI enters the picture, and changes the calculation.

When the path from here to there is murky, AI makes it explorable. You can test possibilities without commitment, map routes you couldn't have charted alone, surface questions you didn't know to ask. Instead of staring at a blank page wondering what your strategy should be, you can say: "Act as a panel of experts across marketing, operations, and finance. Ask me every question you would need answered to understand my situation. One at a time." And then you let it interrogate you. Not to generate an answer, but to sharpen the problem. Or you can say: "Before we build anything, reflect back what you understand about my goals and constraints. What doesn't make sense? What am I missing?" Context is the fuel for insight—that's as true here as it was in our earlier exploration. The more context you give it, the better it performs. The goal isn't to *outsource* thinking. It's to create the conditions for *better* thinking.

The same shift happens with execution. When implementation

looks impossible—too much to do, already stretched—AI doesn't do (all of) the work for you, but it can multiply your own capability. It can decompose overwhelming tasks, draft what you'll refine, research what you'll synthesize, organize what you'll direct. Imagine, for example, that you've clarified your goal, but now you're stuck on the middle 80%—building the webinar, outlining the offer, structuring the curriculum. This is where most momentum dies. Instead of asking AI to "write it for you," you use it as a collaborator. You paste in what you've written so far and say: "Here's what I'm trying to accomplish. Where does this feel structurally weak?" Or: "I'm stuck between two ways of positioning this. Steelman both options." The blank page disappears. The work becomes iterative instead of paralyzing. What once felt out of reach starts to feel within it.

Energy changes too—and this is where people tend to underestimate AI the most. When something knocks you off balance—conflict, setback, overload—your execution collapses. Not because the strategy changed, but because your nervous system did. Instead of spiraling for days, you can open a thread and say: "I need to process something that just happened. Reflect back what you're hearing." First you process emotionally. Then you say: "Now help me think through my options." Then: "Help me outline the next concrete step." What used to take days of cognitive drag becomes one clean arc.

This is what we're recruiting AI for, not a particular tool or technique, but a change in the conditions under which your own thinking operates.

When people first encounter their own expanded thinking, the reaction is often surprising. Sometimes it feels overwhelming. You see the distance between where you are and what now seems possible, and the gap can feel daunting. The gap is real. And it's an honest starting point.

Other times, the reaction is recognition. The sense that this direction, this scale of thinking, feels deeply familiar. Not foreign or inflated, but aligned. As if something that had been waiting in the background has finally stepped forward. What emerges in these moments isn't something new being added. It is capacity being surfaced—capacity that has always been there, operating inside a narrower boundary.

Both reactions—the gap and the recognition—come from the same shift. The capacity was always yours. What changed were the conditions under which it could surface.

For most of our lives, expanded thinking depended on unstable forces: crisis, fantasy, external pressure. Now the conditions themselves can change. The frontier of what feels feasible can expand deliberately. But expanded possibility is not the same as wise action.

If the boundary of the thinkable has moved, the next question is how to navigate the territory responsibly. Seeing a larger horizon is one thing. Choosing the right direction, taking the right steps, and sustaining the effort are another. And expanded feasibility is only the beginning.

In the chapters that follow, we'll look at each of these dimensions in turn—how AI reshapes strategy through disciplined scoping and interrogation, how it transforms execution through iterative collaboration instead of delegation, and how it protects energy by separating emotional processing from decision-making. Expanding imagination is powerful. Learning to work inside that expanded field is what makes it durable.

I know because in May of 2025, everything I'd been building was tested at once.

My father-in-law was dying. A major launch was underperforming badly. The business was under real financial pressure, the kind where you check the numbers every morning and the story doesn't change. And I was the only person holding the full picture. My wife was grieving her father. My team was executing the launch. Nobody else could see all the pieces at once, and there was no one I could hand that weight to.

The thought that kept coming back, in different forms and at different hours, was simple: there's nobody else. It has to be me.

Over the weeks that followed, I had to apply everything you're about to read—strategy under pressure, execution when the plan is falling apart, energy management when there's nothing left in the tank—under conditions I wouldn't wish on anyone. And somewhere in the middle of it, while I was using AI to think through problems I couldn't see clearly on my own, I realized that what I was doing with these tools was unlike anything I'd seen anyone else teach. The approach that was holding me together during the hardest stretch of my professional life was also, it turned out, something worth sharing. That realization eventually became this book.

Not because my experience is the point, but because these ideas only matter if they hold up when things get hard. They did.

Before You Move On

This chapter argued that the biggest constraint on what you can achieve isn't courage or creativity—it's your sense of what's feasible, which operates invisibly and determines which ideas even form in the first place. Key ideas:

- We all operate inside a feasibility frontier shaped by experience, resources, and what we've seen work and fail. Ideas that fall outside that frontier don't get rejected—they never arise.
- What feels like "common sense" about what's possible is often just today's constraints treated as permanent truths.

- Three structural dimensions define the frontier: strategy (can you see a path?), execution (can you walk it?), and energy (can you sustain it?). When any one feels out of reach, the idea doesn't surface.
- AI changes the conditions themselves—making strategy explorable, execution less overwhelming, and energy recovery faster—without crisis, fantasy, or external pressure.
- The capacity for bigger thinking was always yours. The ideas just need room to form.

CHAPTER 6

Strategy in an AI-Amplified World

ONE OF THE hardest things a leader can do is walk away from something that's working. Something still generating revenue, still serving customers, still doing exactly what it was designed to do. Usually, the hardest strategic decisions aren't to fix what's wrong. They're to abandon what's right, because you can see that "right" has an expiration date.

For instance, in 1985, Intel was a memory chip company. They'd invented DRAM, and it was their identity. But Japanese manufacturers had eaten their market share, and Andy Grove asked his co-founder Gordon Moore: "If we got kicked out and the board brought in a new CEO, what would he do?" Moore didn't hesitate. He'd get out of memory chips. So they did, and went on to dominate computing for three decades with microprocessors.

Apple killed the iPod at the height of its revenue to launch the iPhone. Billie Jean King left the established tennis tour, to co-found an independent women's organization—risking her career on something that didn't exist yet. Arianna Huffington walked away from a media company she'd built to nine figures to start something entirely different. Netflix let a billion-dollar DVD-by-mail business die because they could see streaming was where things were heading.

I did something similar when I retired our entire product catalog.

Not one program—all of them. Fifteen years of online courses for experts and entrepreneurs, still selling well, discontinued. I did it because I saw that the market was shifting. AI was transforming how people learn, the traditional course model was declining, and I'd come to believe that online courses simply weren't the best starting point for most of the people we serve. The products no longer reflected how things actually work. So I pulled the plug.

Would I recommend the same move for everyone? No. It depends entirely on what you're trying to build and whether the thing you're walking away from is actually in the way. For me, it was. The old catalog was consuming resources that needed to go somewhere else, and once I saw that clearly, the decision was straightforward. That's what strategy does. It strips away everything that doesn't matter. What's left is a plan of action designed to achieve a goal.

A good strategy can be explained in a sentence or two, but most people and most businesses can't do that. Instead of choosing, they keep busy doing "all the things" and hoping it all clicks. They confuse activity with direction, generate motion instead of strategy, and the calendar stays full while nothing changes. There's always another project to start, another initiative to explore, another idea worth considering. Because everything feels productive, nobody stops to ask whether any of it connects to a decision about what actually matters.

And AI, used poorly, makes this much worse. It's extraordinarily good at generating options, and generating options feels productive. You ask for ideas and get twenty. You ask it to elaborate and get a detailed plan for each one. Before long you have more possibilities than you started with, more frameworks to consider, more directions you could theoretically pursue. None of it is wrong, exactly. But none of it helps you decide. And decision is what strategy actually requires. The danger isn't just that AI gives you bad advice. It's that it gives you so much plausible-sounding advice that the hard work of choosing gets buried under the sheer volume of output.

But AI can also help with strategy, if you bring it into the process the right way. It can pressure-test your assumptions, surface constraints you hadn't noticed, and help you think through the implications of a choice before you commit. This is the Clarity Cascade applied to strategic decisions—the same progressive reorganization we explored earlier, but focused on a specific question: what are you going to do, and why? The key is the sequence. Get clear on what you actually want. Translate that into a goal specific enough to filter decisions. Find the one constraint that's actually in the way. Build your plan around breaking it. AI can support each step, but the order matters.

WHAT YOU ACTUALLY WANT

The prerequisite for strategy is knowing what you want. Not at the level of targets or metrics. This is the step before that.

What kind of work do you want to spend your days doing, and what kind of life do you want your business to support? Do you want to build a team, or would you rather stay lean and close to the work? What are you unwilling to sacrifice, even if it means growing slower? These are questions almost nobody sits with long enough to answer honestly, because we're surrounded by signals about what we *should* want: more revenue, a bigger audience, scale. These are the desires articulated at conferences, modeled by peers, and celebrated in every success story on your social feed. And with enough exposure (and not enough introspection), they become indistinguishable from your own ambitions.

None of these aspirations are wrong. Revenue matters. Growth matters. But they're default answers, the strategic equivalent of ordering what's popular instead of what you actually want. So you sit down to plan and immediately start talking about redesigning the website, hiring an assistant, launching a new offer. None of it connects to what matters to you, perhaps because you never stopped to ask. AI won't ask for you either. Share a vague aspiration and you'll have a plan by lunch. But this is also where AI earns its keep—not as a planner, but as a mirror.

To get past the defaults, you have to keep asking why. Take whatever target is top of mind, say "increase revenue." Why do you want that? To hire help. Why? To stop doing everything myself. Why does that matter? So I can spend time on what I actually care about. Five rounds in, most people end up somewhere they didn't expect, wanting more time with family, or the freedom to work on problems that actually interest them. The real desire was there all along. It was just buried under what they thought they were supposed to want. And once you've found it, you can see which of your other desires are ancillary, which are nice-to-have, and which ones actually matter.

This is where AI can serve as a thinking partner—not to plan how to get what you want, but to help you excavate what that actually is. You're not asking for strategy yet. You're asking AI to help you pressure-test the desire itself, to surface what's borrowed and what's real.

Your AI Power Move: Clarify Desire

Share what you think you want and ask AI to reflect it back in one sentence. Does it sound like you, or like someone else's ambition? Then keep going. Surface the contradictions—you say you want freedom, but the plan requires controlling every detail. You say you want deep creative work, but the target only works at scale. Ask what you'd have to stop doing if this desire were real. What would success look like on an ordinary Tuesday, not the highlight reel?

The conversation usually takes a few turns. You share what you want, AI reflects it back, and you realize you need to adjust. You adjust, it reflects again, and each round gets a little closer to something that actually sounds like you. By the time you've gone back and forth three or four times, the desire you're looking at is simpler and more specific than the one you started with, and harder to dismiss. Rank your desires against each other—which are core, which are ancillary, which are nice-to-have. The ranking is yours, not AI's. That's the one you build on.

The point isn't to find a perfect answer. You're looking for an honest one, a desire you own rather than one you borrowed. Without that, every plan you make optimizes for someone else's priorities. And while desire tells you what matters, it doesn't tell you what to do next. For that, you need the right goal.

GOALS THAT FILTER

Once you're clear on what you want, the next step is translating that into a goal specific enough to act on. This is where most people stall, because choosing a goal means closing doors.

The common assumption is that goals are about motivation—something to aim for, something to keep you going. But a useful goal is not a motivational statement. It's a constraint. Psychologist C.R. Snyder called this Pathways Thinking—when you define a goal clearly, your mind orients toward it. It generates specific routes that could lead you to the goal. And (since time and energy are finite) your mind evaluates which ones could actually work. The rest fall away. So a goal that forces a tradeoff is a goal that filters.

Say your goal is to "grow the business." You could launch a podcast, run ads, write a book, hire a salesperson, build a course, partner with influencers. All of them could work. And because any path is a reasonable path, there's no basis for choosing. That's the problem with a vague goal. It doesn't filter anything. But change the goal to "double revenue from existing clients by December," and most of those options fall away overnight. A podcast won't do it. Neither will the book. Maybe two or three strategies on the entire list could actually get you there. The goal didn't generate new ideas. It eliminated the ones that couldn't work.

And the filtering only works if the goal is calibrated correctly—specific enough to eliminate most options, ambitious enough to force tradeoffs. If it's too vague, nearly anything qualifies. If it's too modest, nothing meaningful has to change. Benjamin Hardy, building on Snyder's work, argues for aggressive goals—because goals that stretch you clarify thinking, precisely because they make tradeoffs unavoidable. The more ambitious the goal, the fewer paths survive, and the ones that remain become obvious. Once a goal is clear, most strategies disqualify themselves.

A quick test is to add "so that" after your goal. Grow my email list *so that* I can launch my course *so that* I can generate enough revenue *so that* I can . . . If the chain keeps going, you haven't arrived at the real goal yet. A goal that's really a step toward something else can't filter—it leaves too many paths open. The real goal is the one where the chain stops. This is where AI helps, not by generating strategies for a vague goal, but by pressure-testing the goal itself.

Your AI Power Move: Stress-Test the Goal

Run the "so that" test. Take your goal and keep adding "so that" after it. Where does the chain end? If it keeps going, you haven't found the real goal yet. Then ask AI to run a pre-mortem: if you achieved this goal and still felt dissatisfied, why would that be?

A few rounds in, the goal is either sharper than when you started, or you've realized it wasn't the real goal at all. Either way, you decide—commit to the sharper version, or go back and find the real one.

A goal that's doing its job eliminates more than it keeps. If yours isn't narrowing your options to the few that could actually work, it isn't specific enough yet. Then once you have a clear goal, the next question isn't where you're going but what's standing in the way.

THE ONE THING IN THE WAY

Most people jump straight from goal to planning—brainstorming tactics, mapping out timelines, building systems. But there's a step between having a goal and knowing what to do about it, and skipping it is the most common strategic mistake I see. It's easy to make, because the work people choose instead isn't wrong exactly. It's logical, it's productive, it's clearly connected to the goal. But as the saying goes, the definition of unproductive is working on the second most important thing. You have to find the *one* thing that's actually standing between you and the goal.

Eli Goldratt made this point in *The Goal* by showing that in any system, there's only one constraint that actually limits everything else, the same way there's only one narrowest point in a bottle. Work on anything that isn't the bottleneck and you're wasting effort. A sandwich shop with a long line at the register and nobody waiting for sandwiches doesn't need another sandwich maker. It needs another register.

Ask an entrepreneur what they're working on and you'll hear five things at once. Grow sales, grow the audience, launch the book, redo the website, take more vacation. They're working on all of it, making incremental progress on none of it, and feeling busy the entire time. That's what happens when you haven't identified the bottleneck. Without one, there's no basis for choosing what to work on first, so everything stays on the list and nothing gets the focused attention it needs.

Almost nobody names a real bottleneck when you ask them. They say time, money, or "me." And those answers feel true, because the pressure behind them is real. There's never enough time. The money never stretches as far as you want. And people who want to make an impact feel responsible for everything. But a real bottleneck is a specific constraint, one that limits the entire system and that you could start working on tomorrow morning. Time, money, and "me" are signals pointing at something deeper.

The way to find the real constraint is to trace those signals backward until they resolve into something concrete. If you think money is the bottleneck, ask why there isn't enough revenue. Because you don't have enough customers. Why? You've relied entirely on referrals and they've slowed down. Why? You never built an acquisition channel beyond word of mouth. Why? Because you've never figured out how to describe what you do in a way that resonates with people who don't already know you. Now you're somewhere useful. The bottleneck isn't "money." It's a messaging problem. That's specific enough to work on tomorrow morning.

The temptation is to ask AI "what's holding my business back?" and let it diagnose. It will, and you'll get a polished breakdown of five or six interconnected factors, each one plausible, none of them the bottleneck. Complexity is not the same as specificity. Don't ask AI to diagnose. Use it to trace.

Your AI Power Move: Trace the Constraint

Think of this as a diagnostic conversation, not a single question. Share your symptoms—what's not working, what you've tried, what the pattern looks like from the inside—and ask AI to describe what it sees from the outside. Then have it name three candidate constraints and eliminate the ones the evidence doesn't support. Which one actually explains the pattern? Which single fix this quarter would create cascading improvement across the others? Then ask yourself: is this the constraint you're used to naming, or the one that's actually limiting progress? Name the real one.

The diagnosis is yours, not AI's.

The goal isn't for AI to hand you a diagnosis. It's for the conversation to surface something you hadn't seen clearly on your own. Once it does, you name it.

The constraint does what the goal does. It narrows. Instead of everything that could be wrong, you have one thing that is. And once you have that, the question of what to do about it becomes surprisingly answerable.

FROM CONSTRAINT TO COMMITMENT

Once you've named the real constraint, strategy becomes surprisingly clear. Not easy, but clear. Most people assume the hard part of strategy is figuring out how to solve their problem. But identifying what the real problem is matters more. After you've accurately named the bottleneck, the "how" becomes a solvable question. You can research it, hire for it, ask someone who's done it before. Sometimes the bottleneck you identify is itself "learn how to do X," which sounds like a how but it's actually a what. You've named a specific capability gap, and that's a bottleneck you can work on.

Strategy is the plan for breaking that bottleneck. Not "what should I do?" but "what's the best way to address this constraint, given these goals, in service of what I actually want?" That's a question with a manageable number of answers. And that narrowing is the point. Not a bigger whiteboard with more ideas, but a smaller space with one clear bet. Most people never get here because they start from the wrong end. They start with a tactic, a new funnel or a rebranding or a product launch, and justify it as strategy after the fact. If you haven't clarified your desires, sharpened your goal, and named your bottleneck, no amount of planning will compensate.

You can hear the difference in how someone enters the conversation with AI. Not "what's the best way to restructure my offer?" but "I think the problem is my offer. Six months of traffic, almost no conversions. I've tested three price points and two audiences. The only thing I haven't changed is the offer itself. What am I not seeing?" The first question hands AI a blank page. The second gives it a specific problem with real constraints. That's someone who's done the thinking and is using AI to sharpen it.

You've built something worth defending at this point. A clear desire, a filtering goal, a named constraint, and a plan for breaking it. The last step is to stress-test that plan before you commit.

I described earlier in this chapter how I retired our entire product catalog. What I didn't mention was what happened next.

The launch webinar was the strongest I'd ever created, the delivery was excellent, and the audience was engaged. And yet, the new launch landed badly. Not a little below target—more like 75% down from what we needed. It was a great launch, and the market said no. I delivered the best work of my career, and it wasn't enough.

After the dust settled, I sat down with AI to diagnose what had gone wrong. I wasn't looking for comfort or reassurance. I wanted to think clearly about a problem that I was too close to see on my own. Context is the fuel for insight, and I had plenty of context—just no distance from it. I brought the 40/40/20 rule to the conversation—the principle that 40% of an offer's success comes from the right product, 40% from the right audience, and 20% from the marketing. The marketing had tactical missteps, including a confusing price anchor in the copy. But as AI helped me trace the pattern, the deeper issue came into focus: offer-audience fit.

The offer required people to do something they were psychologically resistant to. It asked them to reach out to people they already knew, make direct offers, be visible in ways that felt uncomfortable. It was the right methodology—our best client results had come from exactly this approach, and the people who followed through got extraordinary outcomes. But it required an emotional stretch that most buyers weren't ready for at the point of purchase. They were hanging on my every word during the webinars because the ideas made sense. I could see it in the engagement, in the testimonials, in the questions they asked. But making sense and being ready to act are different things.

So I mapped out the options. Door one: sell people the comfortable fantasy, the automagic formula that generates results on autopilot. That's a common playbook in my industry—it would convert, but it's not real and it doesn't work. Obviously, that approach is completely out of integrity, so I wasn't going to do that. Door two: double down on the same message, insist they do it the "right" (i.e. my) way. That's pushing a boulder uphill, and it ignores what the data was actually telling me. Door three: reorder the funnel so people enter through a less confronting offer and graduate into the harder work when they're ready for it. Door four: build a foundational entry point that develops the beliefs and confidence people need before either main program becomes the obvious next step.

This was the Clarity Cascade applied to a real strategic crisis, under real pressure, with real money on the line. Each pass through the question reorganized the problem. The frame shifted from "the launch failed" to "we were solving the right problem in the wrong sequence." The constraint wasn't the content or the delivery or the audience. It was the order in which people encountered the ask. Once I could see that, the path forward was clear—not easy or comfortable, but clear. And that clarity came from the same process this chapter describes: staying in the conversation long enough for the real constraint to surface.

Your AI Power Move: Challenge the Strategy

This is where you find out whether the plan holds up. The goal here isn't to get AI's approval. It's to find the gaps in your own thinking before you commit resources to closing them.

- **Reflect:** Share your full logic chain with AI—your desire, goal, constraint, and plan. Ask it to reflect back what it heard.
- **Pressure-test:** Ask AI to identify the load-bearing assumptions, the ones that would collapse the plan if they turned out to be wrong. What dependencies haven't you tested?
- **Sequence:** Ask for ordering, not more ideas. What do you do first, and what does it cost?
- **Decide:** Does the plan hold, or did something just change? The bet is yours, not AI's.

The conversation tends to sharpen rather than expand. You lay out your reasoning, AI pushes back on an assumption or surfaces a dependency you hadn't considered, and you either adjust the plan or confirm why your original logic holds. Each round makes the bet cleaner. When the plan holds, you commit.

The plan that emerges from this process is leaner than what you started with. Fewer assumptions, fewer dependencies, a single bet you've tested before committing. A real strategy is a plan for breaking the one constraint that matters most. If the plan survives this conversation largely intact, that's a good sign. Not because AI validated it, but because you were able to defend it under pressure. And if it didn't survive—if AI surfaced something that changed your thinking—then the conversation just saved you months of executing the wrong plan.

THE MEASURE OF REAL STRATEGY

You know the process has worked when it produces less, not more. One desire you're willing to own. One goal specific enough to filter decisions. One constraint you can name. One bet you're prepared to make. That's it. Not a sprawling strategic plan, not a vision board, not a list of everything you could do. Just the clearest possible answer to the question: what am I going to do, and why?

This is what most people get wrong about strategy, and what AI, left to its own devices, will get wrong too. We tend to associate strategic thinking with expansion, with seeing more possibilities, generating more options, covering more ground. But the work of strategy is compression. You start with everything you could want and end with the one thing you're going to pursue. You start with every possible obstacle and end with the one that actually matters. You start with a dozen ways forward and end with a single bet. If the output of your strategic process is a longer list than you started with, something went sideways.

The hard part isn't the thinking. It's the accepting. Every real strategy carries a cost—the projects you won't pursue, the opportunities you'll walk past, the version of your business you're choosing not to build. AI can help you think through the decision, but it cannot carry the weight of it for you. That weight is what makes it real.

Choosing a direction is only half the work. Strategy tells you where you're going and what you're willing to bet on. But plenty of clear-eyed plans have died in the gap between decision and delivery—not because the strategy was wrong, but because sustained daily action is a different kind of thinking entirely. That's where execution comes in.

Before You Move On

This chapter applied the Clarity Cascade to strategic decisions, working through a four-step sequence: clarify what you actually want, translate it into a goal specific enough to filter decisions, identify the one constraint that's actually in the way, and build your plan around breaking it. Key ideas:

- The prerequisite for strategy is knowing what you want—not at the level of targets, but at the level of desire. Most people optimize for borrowed ambitions without realizing it.
- A useful goal is a constraint, not a motivational statement. The more ambitious and specific the goal, the more options it eliminates—and that elimination is the point.
- Almost nobody names a real bottleneck. They say time, money, or "me." The real constraint is always more specific, and you find it by tracing those signals backward.
- AI's danger in strategy is generating so many plausible options that the hard work of choosing gets buried. Don't ask AI to diagnose. Use it to trace.
- Strategy produces less, not more. One desire, one goal, one constraint, one bet.

CHAPTER 7

Collapsing the Execution Gap

EVERYONE HAS A GRAVEYARD of good intentions. Plans that made sense on paper but stalled somewhere in the middle, ideas that seemed promising but never became a finished product. The effort and the drive were real, but something between the idea and the result kept falling apart. The previous chapter was about using AI to find a direction worth pursuing. But a clear direction doesn't automatically become a finished result, and the reason is rarely what people think.

Most advice on execution lands in one of two places. Try harder—be more disciplined, show up earlier, stay later, push through the resistance. Or build better systems—habits, checklists, accountability structures that force you to do the right things whether you feel like it or not. One assumes it's an effort problem. The other assumes it's an organization problem. But if either were the real bottleneck, the most dedicated people would have the fewest stalled projects. They don't. Some of the hardest working people I know have the longest list of unfinished work, and the problem isn't laziness or poor systems. They're applying force to ambiguity. Working harder when the work itself isn't clear enough to build produces exhaustion, not results. And the best-designed system in the world can't save you from running the wrong process with great consistency. The ambiguity isn't always obvious, which is why the temptation to just hand it all over to AI is so strong. But you can't delegate ambiguity. You have to resolve it, and you can't resolve what you don't see.

When my daughter Priya was 8 or 9 years old, she was very into building Harry Potter Lego sets—the big ones that say 18+ on the box and have thousands of pieces. She was patient and followed the instructions, sometimes spending weeks on a single build. But every so often she'd stall, and discover a mistake from a hundred and fifty steps ago. The piece didn't fit. The wall didn't align. And she'd just flop down on the couch, face first, totally defeated. The project wasn't ruined, but she knew what had to come next: taking apart hours (and sometimes days) of careful work to get back to one wrong piece. We all know the advice to measure twice, cut once. But measuring twice can't save you from building the wrong thing with great precision. This is the illusion of clarity—the sense that we understand something far better than we actually do. And if clarity isn't where you think it is, the question becomes: where does it actually come from?

It doesn't arrive *before* the work starts. It emerges *through* the work. Joan Didion said, "I don't know what I think until I write it down." You discover what you actually understand by trying to make it real. This is the Clarity Cascade in a different key—not applied to a strategic question, but to the work itself. Each pass through the work surfaces what the previous pass hid, and understanding reorganizes as you go. You decide what you're building, start building, and that process reveals that your understanding was incomplete. You refine what you've built and discover the structure needs rethinking. So you go back, and you keep going back—designing, building, polishing, and then discovering the design needs to change again—because each pass exposes what the previous one hid. Those three phases are the core loop of execution. You design to surface what you're actually building, build through cheap iteration that lets you discard and restart without flinching, and polish to find what breaks before your audience does. AI changes the cost of every pass through that loop, and that changes more than you'd expect.

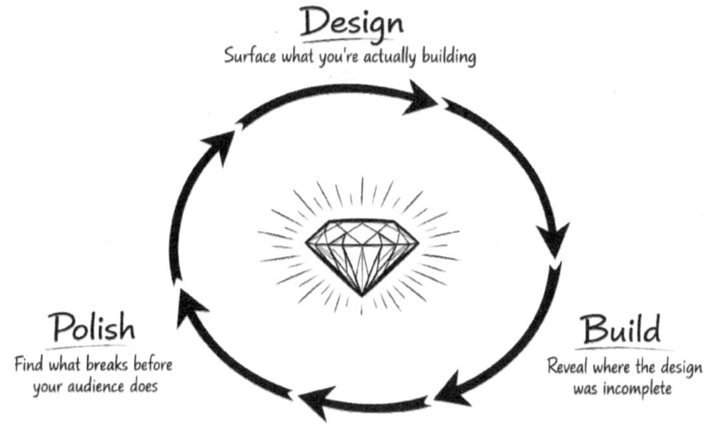

DESIGN: WHAT YOU'RE ACTUALLY BUILDING

Rushing through design is tempting because it doesn't feel like productive work. You want to get moving, start producing, and see it take shape. But the gap between what feels clear and what you can actually articulate is almost always larger than you expect. Say you're developing a new program and you sit down to write the first module. You've got a dozen open questions—who this is actually for, what they already know, whether to include exercises, what the real transformation is. So you start writing anyway, hoping clarity will catch up as you go. And sometimes (more often than anyone admits) that momentum carries you straight into a wall you could have seen coming.

What works better is emptying everything out before you ask for anything. Your thinking, your doubts, your half-formed ideas, the constraints you're working within—all of it, in whatever order it comes out. I sometimes tell AI not to respond until I say I'm done, just to keep the brain dump going without stopping to evaluate. The specific words don't matter. What matters is that you externalize your context before you ask for solutions. Context is the fuel for insight, and design is where you load the fuel.

This is what my weeks look like now. An idea shows up—a new workshop format, a way to restructure our onboarding, a partnership that might work. I get everything out, and then I ask for an outline. What used to take a week of working it out on my own takes a few minutes. And what comes back is almost never right, because the thinking behind it wasn't sharp enough yet. But when a rough draft costs almost nothing, that's fine. If I'd handed the idea to someone on my team, they'd have put real hours into it, and now I'm in the position of explaining that the work isn't usable—not because they did anything wrong, but because I hadn't thought it through. With AI, there's no sunk cost and no one to let down. I look at what came back, see where my thinking was off, and go again.

Michelangelo said he saw the angel in the marble and carved until he set him free—but most of us don't know what the angel looks like until we've done quite a bit of carving! That's what the rough draft is for. Not to use, but to hold up against what you meant. Read it not to evaluate the quality of AI's writing but to evaluate the quality of your design. If the prototype surprises you, the design needs more work. If it feels recognizable, you're ready to build. You don't design to avoid iteration. You design to make each iteration count.

Your AI Power Move: Design Before You Build

Before you ask AI to produce anything, dump everything in your head first—your thinking, your doubts, your constraints, in whatever order they come. Then ask for a rough draft. Not to use, but to hold up against what you meant. If the draft surprises you, your design needs more work. If it feels recognizable, you're ready to build.

The prototype almost always surprises you. Not because AI got it wrong, but because seeing your thinking externalized reveals gaps you couldn't see from the inside. You thought the scope was clear, but the summary includes something you didn't intend. You thought you knew the constraints, but the assumptions list includes one you hadn't stated. A few rounds of this and the design tightens in ways that feel obvious in hindsight. When it stops surprising you, you're ready to build.

Every assumption you don't surface in design becomes a surprise in the build. The prototype exists to find those surprises while they're still cheap to fix. Design tells you what you're making. But knowing what you're making and actually making it are different kinds of work, and the second one is where most people lose their way.

BUILD: THE MESSY MIDDLE

Building is the middle 80 percent. Drafting, developing, iterating, producing. The pace is uneven. Some hours the work pours out, and then you hit a wall and nothing comes at all. This isn't a sign of failure. It's what building looks like. And the quality of what you produce depends not on how long you spend but on how many times you iterate. A first draft that took a week isn't necessarily better than one that took two hours—especially if the week was mostly going down blind alleys. Real iteration means you actually changed something between versions; you questioned an assumption, reorganized a structure, cut something that wasn't working. Without that, you're just going in circles. And it's surprisingly hard to tell which one you're doing from the inside.

When people talk about AI and execution, the conversation usually jumps straight to automation and agents—handing whole tasks to AI and letting it produce the finished result. And if you do it right, that approach can be very powerful, which is why we'll get to that in the next chapter. But that works best when you've already developed something more fundamental: the ability to use AI to sharpen each pass through the work while you're the one doing it.

And the pass that matters most is the one where you realize something isn't working. You've reorganized the structure, refined the language, spent days getting the pieces to fit. And now you can see that the foundation underneath it all needs to change. Without AI, that realization is devastating. You've invested too much to start over, so you patch. You work around the weak spot, add transitions to cover the seam, convince yourself it's close enough. With AI, rebuilding a section costs minutes, not days. You can pull apart the middle of a project and reconstruct it without losing a week of momentum. The question stops being "can I afford to redo this?" and becomes "is this actually working?" That's a different kind of honesty, and it produces different work.

This played out for me during that fateful May launch, when the stakes couldn't have been higher.

The first webinar of the launch was the day after my father-in-law had passed away. My delivery was competent—the audience responded well, my team said it was good, and plenty of complimentary testimonials came in. But I knew it wasn't what it needed to be. The timing was off, and I didn't feel the congruence between the stakes of the moment and the quality of the delivery. I know when I nail it, and that day I didn't. I came home exhausted, grieving, and disappointed in a way that was hard to explain to anyone who'd watched the same webinar and thought it was fine. I felt like what the business needed was a home run, and what I'd delivered was a solid double at best.

I woke up the next morning drained. But I had another delivery that afternoon, and I wasn't going to let the second one go the way the first had. I spent the morning revising the webinar with AI as my collaborator—not having it rewrite the content, but using it to identify where the structure was weak and where the argument wasn't landing the way I wanted it to. We went section by section, and each pass tightened something I hadn't been able to see from inside the delivery. I finished the edits twenty minutes before going live, feeling a lot of trepidation about delivering a three-hour training on that little sleep and that much emotional weight.

That second delivery was what the first one should have been. The engagement was stronger, the pacing was right, the content landed the way I'd designed it to land. That's the design/build/polish loop under real pressure. The first delivery surfaced what wasn't working. The morning revision was the build phase compressed into a few hours, because AI made iteration nearly free. And then I tested it live—not a rehearsal, but the real thing, with a real audience.

Nobody else saw the gap between the two. But I could feel it, and that gap—between how hard something is to pull off and how casually it gets received—is its own kind of weight. I'll come back to that. But the work got done, and it got done because I wasn't doing it alone.

When you get stuck, AI can function as a thinking partner—not a suggestion generator, but something closer to a colleague who helps you work through the problem out loud. You're mapping out a pricing model and the tiers don't make sense together, so you lay out the logic and ask AI to identify where the structure breaks down. You're developing a workshop and the opening exercise feels flat, so you describe what you need and ask for five different approaches. Most of what comes back won't be usable, but one suggestion points somewhere you hadn't considered. You don't use AI's version. You write your own, informed by where it pointed. That's the collaboration—not AI producing the work, but AI helping you see it differently. The more fluent this becomes, the more natural it feels to hand AI not just the stuck points but the work itself.

Your AI Power Move: Build Through Iteration

The next time you realize something isn't working—a section that's holding together on momentum rather than merit—try throwing it away on purpose. Ask AI what you'd lose if you cut the weakest part entirely. Does the rest hold up? If it does, you just found dead weight. If it doesn't, you just found the load-bearing piece that needs to be rebuilt, not patched. The first time you do this it feels wasteful. By the third or fourth time, it stops feeling like waste and starts feeling like the process.

Iteration without judgment is just rearranging. Each pass should change what you understand, not just what's on the page. At some point the work holds together. The structure makes sense, the pieces connect, and you're no longer rebuilding from scratch. That's when most people stop. But holding together and being excellent are separated by one more phase, and it's the one almost everyone skips.

POLISH: THE LAST TEN PERCENT

Most people either skip this phase entirely or reach it with too little energy to do it well. They exhaust themselves in the build and call it done when it's merely complete. But the difference between adequate and excellent lives almost entirely in this last 10 percent, the refinement, the edge cases, or whether the structure can withstand real pressure. AI can help here, but the role matters. Ask it to act like an editor who advises rather than one who rewrites for you. "Here's the project plan I'm about to share with my team. It feels pretty robust to me, but I'm sure I have blind spots. What are they?" The question isn't "fix this." It's "help me see what I'm missing." One makes you better. The other makes you dependent.

The hardest part of the polishing phase is objectivity. By this point you're deep inside the work. You know what you meant, and it's hard to separate your intentions from what's actually on the page. You skip past the section that doesn't quite land because you remember the thinking behind it, even when that thinking never made it into the final version. Polish is where you shift from building for yourself to building for the person who'll receive it, and that shift requires a kind of distance that's hard to create after weeks in the weeds. AI can provide that distance. Not because it's a better editor than a human, but because it can read what's there, not what you intended. Ask it to stress-test: where would a skeptic push back? What am I assuming my audience already understands? What's the weakest link in this argument? You'll surface problems you wouldn't have found on your own—not because AI is smarter, but because it isn't invested in being done.

Say you've just finished a proposal for a new partnership. It covers the opportunity, the terms, the mutual benefits, the implementation plan. All the pieces are there, and it's complete. But complete and excellent aren't the same thing. The questions that push toward excellent are specific: does the opening make the case for urgency, or does it ease in too slowly? Does the structure lead to the ask, or does the ask feel tacked on? Will the other side see their priorities reflected, or only yours? AI can surface these questions, but only you can answer them, and only if you know the territory well enough to tell a real problem from a cosmetic one.

Your AI Power Move: Polish Under Pressure

Your audience will find every weakness you missed. Find them first.

- **Check alignment:** Present the finished work and ask AI to summarize what it accomplishes. Does that match your intent?

- **Play the skeptic:** Ask AI to argue against your work. Where does the argument fall apart? What objection haven't you addressed?
- **Test for memorability:** Ask what someone would actually remember a week later. What would they do differently because of this work?
- **Ship or iterate:** Is this ready, or are you calling it done because you're tired of working on it? The answer is yours.

What usually surfaces isn't what you expected. You go in thinking the work is solid, and AI finds a gap in the logic you'd been skating past. Or it catches that your strongest point depends on an assumption you never stated. The objections that matter are rarely the ones you anticipated.

Most weak spots aren't hiding. You just stopped looking before you found them. The difference between work that's finished and work that holds up is almost always one more honest pass, taken when you'd rather be done. Now, in practice, designing, building, and polishing don't stay as separate as they sound. The phases blur.

THE LOOP IN PRACTICE

Say you're redesigning the onboarding process for new hires. In design, you dump everything. What you know about where people get stuck in their first month, the complaints from managers about ramp time, the parts of the current process that are just legacy. As you empty it out, a contradiction surfaces. You want onboarding to be thorough, but you also know that front-loading too much information overwhelms people and slows them down. That tension isn't a problem to resolve. It's the redesign's real thesis. You didn't see it until you got everything out.

Once you get a rough process map, you realize the sequence is wrong. You had compliance training first, but people need context about their role and team before any of that makes sense. So you go back to design briefly, restructure, and start building. Maybe the first week's schedule lands, but the second week drags. Too many required readings, not enough moments where anyone does something real. You describe the problem to AI, get five ways to convert information delivery into active tasks, and one of them—a shadow day with a peer mentor built into week one—is better than what you had. So you rebuild the middle around it. In polish, you stress-test. You ask AI to play a new hire who feels overwhelmed and disoriented. Two transitions between weeks fall apart under pressure. You fix them. You ask what someone would actually remember after their first month, and realize your closing checkpoint doesn't give them a clear picture of what success looks like. You add one.

The onboarding process you end up with isn't the one you started with. The thesis shifted, the sequence reversed, the middle was rebuilt, and the success criteria changed twice. But each version was better because you were building from clearer understanding, not just adding more steps.

When you've been around the loop enough times, you stop treating the first version as the real thing and everything after it as corrections. The first version becomes what it always was, a way of finding out what you're making. The second is where you start building with your eyes open. And somewhere around the third or fourth pass, if you've been honest with yourself at each stage, the work starts holding together in ways that surprise you. AI didn't do the work, but the loop gave you room to do yours.

WHEN SPEED WORKS AGAINST YOU

The danger is that AI makes it easy to skip the loop entirely. I was interviewed on a guy's podcast recently, and he was telling me about the system of AI agents he'd built. Every podcast episode they recorded was going to be automatically turned into a mini-course. No human curation, no redesign for a different format, just automated transformation. And I remember sitting there thinking, "but who wants that?" The episodes weren't designed to be courses. The audience didn't ask for courses. The only reason to do it was because the technology made it possible at basically no cost. But if it's not worth spending some time and money doing it, it probably isn't worth having it done for free either.

AI can generate content, plans, and deliverables at a speed that feels like progress. But producing more doesn't mean producing better, and it's easy to mistake a full pipeline for a functioning one. The loop is the antidote—each phase forces a question that pure volume never asks. Is this the right thing? Is it good enough? Does it serve the person it's meant to serve? Those questions only work if you can answer them, and answering them requires real understanding of the work. AI can accelerate work you understand, but it can't replace understanding itself.

Once you can run the loop fluently—injecting real judgment at every stage—a new question emerges. How much of the work can you hand to AI, and how much do you need to hold? The answer depends on how well you understand what good looks like, which is exactly what the loop develops. But knowing where the line is and knowing when to move it are two different skills.

Before You Move On

This chapter argued that the gap between a clear plan and a finished result isn't an effort problem or an organization problem—it's an ambiguity problem. Clarity doesn't arrive before the work starts. It emerges through the work, in a loop of designing, building, and polishing. Key ideas:

- You design to surface what you're actually building, build through cheap iteration that lets you discard without flinching, and polish to find what breaks before your audience does.
- AI changes the cost of every pass through the loop. Rebuilding a section costs minutes, not days—which changes the question from "can I afford to redo this?" to "is this actually working?"
- The design phase is about externalizing your context before you ask for solutions. Context is the fuel for insight, and design is where you load the fuel.
- The polishing phase is where you shift from building for yourself to building for the person who'll receive it. AI provides distance because it reads what's there, not what you intended.
- The danger is that AI makes it easy to skip the loop entirely—producing volume that feels like progress but serves no one.

CHAPTER 8

Earned Leverage: Working with AI Agents

THE BIGGEST FANTASY in AI right now isn't about getting smarter or more creative. It's about building an army of AI agents that handles every piece of your business, while you focus on the things only you can do. (Or, if the fantasy goes far enough, while you do nothing at all!)

I hear some version of this almost every week—someone describing systems where agents handle entire workflows, end to end. A new lead comes in, and agents qualify the prospect, draft a personalized outreach sequence, schedule the follow-up, and log the outcome. A new employee joins, and agents handle the onboarding, schedule the training, generate a development plan. Every customer inquiry routes through an agent that triages, drafts a response, escalates when needed, and closes the loop automatically. The specifics vary, but the dream is the same. Build it once, and let it run forever.

The appeal is understandable. If AI can help you think more clearly and produce better work, the next question is obvious. Why not build systems that handle the production entirely?

It's not impossible—the technology is real and improving fast, and some truly remarkable things are being built. But most people who try to build agent systems fail, and the failures aren't usually technology problems. The agents work fine. The automations run. The output appears on schedule. What doesn't work is the output itself, because nobody with real understanding was directing it, and so everything that gets produced—efficiently, professionally, and delivered on time—doesn't actually serve anyone. This isn't new. The same pattern played out with outsourced talent, virtual assistants, and every previous generation of automation. The technology changes, but the mistake doesn't.

WHAT AGENTS ACTUALLY ARE

Everything in this book so far has described one interaction model: you and AI, thinking together. You give a prompt, AI gives a response, and everything comes back to you. It's a back and forth between you and the tool. No matter how sophisticated the conversation gets, you're always at the center of it, directing, evaluating, deciding what to do next.

Agents work differently. With an agent, AI itself is at the center of the interaction. It can be triggered by a person, the way you're used to working with AI, but it can also be triggered by a system or by another agent. And instead of just producing responses, it can act in the world—on people, on systems, and on other agents. That last part is what creates disproportionate leverage. When agents can talk to each other, trigger each other, and coordinate with each other, the possibilities expand in ways that a single back-and-forth conversation never could.

The same principle that makes specificity so important in prompting applies to agent design too. Narrow, specialized agents that do one thing well consistently outperform a single broad agent trying to handle everything. This means that one agent can't simply replace one person's job. The work is too varied, too context-dependent. In practice, what a single human does might require five, ten, or even twenty-five specialized agents working together—either daisy-chained like a relay race, where one hands off to the next, or organized around a central orchestrator that calls on specialists as needed. Think of it less like hiring one brilliant generalist and more like assembling a team of specialists, or what some people are calling agent swarms.

We've been building something at Mirasee—an AI advisor tool that handles an initial intake coaching conversation. Just the first thirty minutes or so of what a human coach would do. And even getting that right required a team of eight specialized agents working together. One interprets the meaning and emotion behind each message. Another tracks the conversation's progress, what topics have been covered, what insights have surfaced, where the diagnostic stands. A third decides what the next conversational move should be. Others construct the actual instructions for the language model, generate the response, run quality checks before anything gets sent, extract and structure insights from the conversation, and manage the transition when the session needs to close. Eight agents, just for a thirty-minute conversation. And that's what it takes to do one narrow thing well.

The difference between conversational AI and agents isn't just a matter of scale. It's a different interaction model. In a conversation, you're steering. With an agent, you're launching. But the underlying skills that make either one work—clear thinking, good judgment, the ability to define what you actually want—are the same. It's just that with agents, the stakes compound. In a conversation, you catch a wrong turn in real time and adjust. With agents, a wrong turn early becomes the foundation for dozens of downstream decisions, each one logical on its own, each one building on a direction that was off from the start. The same skills, but a different kind of precision, because mistakes don't wait for you to notice them.

WHERE YOU GRIP THE HANDLE

Think about how you hold a hammer. Grip it near the head and you have precise control—you can place a nail exactly where you want it, tap gently, direct every movement. But there's almost no power behind the swing. Grip it at the very end of the handle and the opposite is true. Plenty of power! But good luck placing that nail with any accuracy. You'll bend it, dent the wood, drive the whole thing crooked—and now you're pulling nails instead of building. (There's also the risk to your thumb . . .)

The useful spot is somewhere in the middle, and where exactly depends on how well you know what you're doing. A beginner grips near the head because that's all they can handle—limited power, but they can see where every swing lands. As your skill develops, you can afford to slide your hand further down the handle, trusting your aim, gaining power without losing control. Nobody starts at the far end. You earn your way there.

AI works the same way. At one end, you control everything; dictating each word, specifying each step, reviewing each micro-decision. Lots of control, but very little leverage. You might as well do it yourself. At the other end, you hand AI a vague objective and walk away. "Build me a marketing strategy." "Create a course from this content." "Handle my customer onboarding." Enormous power to produce output. Almost no ability to direct where it goes.

Agents and agent swarms sit at the far end of the handle. Maximum power, minimum direct control. That's not a design flaw, it's the nature of delegation. You're asking a system to make decisions on your behalf, in sequence, often without checking in. When you have the judgment to set good constraints and evaluate what comes back, that power is transformative. But when your own thinking is vague, or you can't tell good output from bad, the hammer swings you.

What determines whether you can find the right grip? It's a progression. And it matters what order you do it in.

A PROGRESSION, NOT A SHORTCUT

There are three levels to working with AI, and they have to happen in order.

The first is thinking with AI. This is what most of this book has been about. The discipline is more than learning to use AI. It's learning to think clearly enough to direct it. Can you frame what you're actually trying to achieve? Can you name the constraints you're working within, what matters, what doesn't, what you're willing to trade away? Can you set a direction before demanding output? Can you tell when your thinking is clear enough to act on, and when it only feels clear? AI will solve whatever problem you put in front of it, including the wrong one, so the real skill is knowing when you're pointed at the right one. This requires the ability to sit with a problem long enough to know when the thinking is actually done, not just when it feels done. And you can tell when AI is helping you see something new, versus just giving you back a polished version of what you already believe.

The second is applying AI to your expertise. This is where you bring AI into the work itself, iterating on real deliverables in a domain you understand well enough to evaluate what comes back. You direct the process, catch the errors, make the judgment calls. The key requirement at this level is competence. The middle 80 percent of any serious work has to be done by someone who understands it well enough to own the result. And there's a rule I follow without exception: don't ever ask AI to do something if you aren't in a position to assess the quality of what comes back. If you can't evaluate the output, you have no business delegating the production.

The third is augmenting with agents. This is where you delegate execution to AI systems that operate with real autonomy, making decisions, taking actions, producing output across multiple steps without you reviewing each one. This is the most powerful application of AI, and the one that fails most spectacularly when it's reached too early.

Each level builds the capability that the next one requires. You can't direct AI in your work if you haven't first learned to think clearly with it, because you won't know what good direction looks like. You can't delegate to agents if you haven't first developed real competence in the domain, because you won't be able to tell whether agents are producing something valuable or just something that looks valuable. Skipping ahead doesn't save time. It creates fragile systems that look productive until they hit something that requires the judgment you never built.

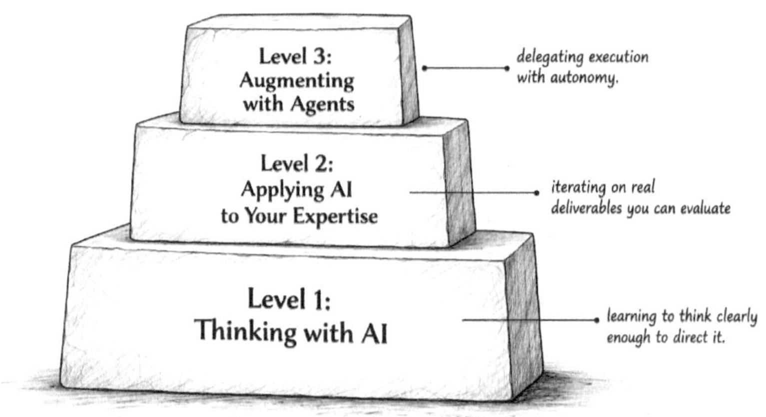

THE COST OF SKIPPING

The failure is usually subtle. Someone gets excited about agents, builds a system that automates a workflow, and discovers weeks or months later that the output was subtly wrong the entire time. Not obviously broken. The grammar was fine, the structure was professional, the deliverables showed up on schedule. But the strategy behind them was off, or the audience wasn't being served, or the work was solving a problem nobody had. By the time anyone noticed, the damage had compounded.

What makes agents particularly dangerous without the foundation isn't just that they produce bad work. It's that they compound it. A regular AI conversation goes wrong in one place. You see the output, catch the problem, adjust. An agent goes wrong early and then makes dozens of decisions on top of that first misstep. Each decision looks reasonable in isolation because it follows logically from the previous step. But if the direction was off from the start, logical consistency just means the system got further off course more efficiently. By the time you check in, you're not looking at one mistake. You're looking at a chain of decisions that all made some sort of sense individually, and collectively produced something useless—or downright damaging!

Without the thinking discipline from the first level, you can't set clear direction. And agents don't question direction. They'll pursue whatever goal you give them, including ambiguous or poorly thought out ones. They don't stop to ask whether the direction makes sense. They don't surface the contradictions in your thinking. They just execute.

The second level is what gives you the ability to tell whether agents are producing something good or something that merely looks good. AI is excellent at producing work that passes the surface check. Clean formatting, plausible logic, professional tone. But if you don't understand the substance well enough to go deeper, the surface is all you'll ever check. And agents can produce enormous volumes of surface-level quality very quickly.

That speed makes the problem worse. When something isn't working, the natural instinct is to do more of it. More content, more outreach, more automation. Agents make "more" nearly free. So instead of stopping to question the direction, you scale it. The dashboard shows activity. The pipeline is full. But volume was never the goal. Clarity was. And that distinction vanishes when you skip to agents without the foundation to tell the difference.

WHAT THE TRAINING IS REALLY FOR

None of this means agents are bad, or dangerous, or something to avoid. For people who've developed the foundation, agents are transformative. They amplify competence the same way they amplify confusion. The difference is entirely in what's underneath.

Using agents well is an advanced skill, and what makes it advanced isn't the technology—it's the judgment required to direct it. The Clarity Cascade doesn't stop being relevant when you move to agents—it's what built the judgment that makes delegation possible in the first place. The clear thinking you develop through all the in-depth conversations with AI that are described in this book... that's what lets you set directions that agents can follow. And the domain competence you build through ever-deepening iteration is what allows you to tell whether an agent produced something that *is* valuable or something that merely *looks* valuable. Without those, you're not delegating. You're just hoping.

And you don't leave those disciplines behind when you start working with agents. You're still thinking with AI and applying AI to your expertise—just with more leverage and less room to course-correct. Agents add their own demands on top: designing instructions for systems that run without you, decomposing work into pieces an agent can handle, building evaluation criteria before anything runs. But those demands don't replace the foundation. They stack on it.

Your strategy, your execution, your ability to delegate well—it all depends on your capacity to show up with real judgment, day after day. And that capacity is more fragile than most people realize. Not a skill, exactly. More like what powers every skill. When it runs low, even the disciplines you've built start to erode.

Before You Move On

This chapter made the case that agents are the most powerful application of AI—and the one that fails most spectacularly when reached too early. The leverage is real, but it has to be earned through a progression that can't be skipped. Key ideas:

- Agents are a different interaction model. In a conversation, you're steering. With an agent, you're launching. Mistakes compound because agents don't question direction—they just execute.
- The hammer metaphor: grip near the head for control, grip at the end for power. Where you grip depends on how well you understand the work. Nobody starts at the far end.
- Three levels, in order: thinking with AI, applying AI to your expertise, augmenting with agents. Each builds the capability the next one requires.
- The rule that holds at every level: don't ask AI to do something if you aren't in a position to assess the quality of what comes back.
- For people who've built the foundation, agents are transformative. They amplify competence the same way they amplify confusion. The difference is entirely in what's underneath.

CHAPTER 9

AI and Energy: Protecting the Human

STRATEGY TELLS YOU where to go. Execution gets you there. But both run on the same fuel, and it doesn't matter how clear your direction is, or how good your systems are—when that fuel runs low, you stall.

We all know what our best looks like: seeing problems clearly, making decisions without second-guessing, handling complexity without shutting down. And we also know what the opposite feels like—when every email feels like a confrontation, every decision could go wrong, and the simplest task takes three times longer than it should. The difference between those two states is not small—it's an order of magnitude! The gap between what you can accomplish at your best versus when you're depleted is night and day. And your skills, intelligence, or character aren't any different. The only difference is your capacity to use them.

That capacity is what I mean by energy. Not enthusiasm, not mindset, and not the kind of thing you fix by going to bed earlier. Energy is the cognitive and emotional capacity that powers your judgment and your action. In practical terms, energy is the usable capacity to think clearly, tolerate complexity, and act deliberately on what you already know needs to be done.

When energy is full, your strategy is sharp and your execution is precise. When it drains, both degrade together. Your thinking becomes narrow and reactive, reaching for whatever's easiest rather than whatever's right. Your execution turns into busywork. You check things off, but the work that gets done is what's comfortable, not what's important. This isn't an ambiguity problem. You know exactly what needs doing, but you sit down and reorganize your inbox instead, because the important work requires more than you have right now.

We all have a sort of internal battery that powers us through our days. Some days it's full, and we operate at a level that surprises even us. Other days it's nearly empty, and we can't understand how we've ever done anything well. And the real cost isn't just what you lose in the moment. It's what you can't do afterward. You push through the rest of the day on fumes, making decisions you'd normally think through, and the damage compounds without you realizing it.

THE SLOW LEAK

Not all energy loss is dramatic. Some drains don't knock you down. They erode you bit by bit, every day, until you can't remember what full capacity felt like. Here is a non-exhaustive list of the sort of thing that acts as a chronic drain on your energy:

Ambiguity. You're not sure your approach is right. Not convinced it's wrong, but uncertain enough that the doubt follows you into every decision. You can't fully commit to anything because you might need to reverse it, so you hedge, revisit, second-guess. The drain isn't the uncertainty itself. It's the tax it levies on everything else you try to do while uncertain.

Futility. You're working hard but the results aren't showing up. The effort is real, but the feedback loop is broken, and effort without results eventually stops feeling like effort. It starts feeling like waste. You don't stop caring exactly, but the connection frays between what you're doing and why it's supposed to matter.

Winlessness. The gap between where you expected to be and where you actually are keeps widening. You set targets, miss them, adjust, miss again. The problem isn't failure. Failure you can learn from. The problem is that trying and falling short over and over becomes its own kind of evidence—evidence that trying harder isn't enough. That's what erodes. Not your competence, but your belief that effort leads somewhere.

Anxiety. The trajectory itself feels unsustainable. Not a crisis, just a sense that where you're heading isn't going to work, and you can't see how to change course. Your brain responds by running a background threat-monitoring process that never shuts off. Even when nothing is actively wrong, you're spending energy scanning for what might be. The drain is exhaustion from a system that's always on alert.

Each of these creates friction in a different way, and none of them announce themselves the way a crisis does. But they all produce the same effect: the quiet erosion of cognitive bandwidth, draining you slowly enough that you adjust to each new normal without noticing you've adjusted.

What makes them dangerous is that the damage is nearly invisible. You choose tasks that feel productive but don't require real thinking. You postpone the conversation you need to have. You accept "good enough" on work that matters to you. None of these feel like collapse. They feel like being practical, being realistic, managing your bandwidth. But each one is a small concession, and after enough of them, you're operating at a standard you never consciously chose.

AI can interrupt some of these patterns. The Clarity Cascade surfaces what you're actually trying to do, cutting through ambiguity. The execution loop makes the first step startable, which interrupts futility. And a thinking partner breaks the isolation that chronic drains feed on. When ambiguity is the drain, a few rounds of structured thinking with AI can resolve in thirty minutes what would have taken days of circling on your own.

But AI can also accelerate the erosion. Too many options create decision fatigue instead of clarity. Too much output creates the illusion of progress, words generated and tasks completed that don't actually move you forward. The pace AI makes possible can become its own pressure, an expectation to keep up with what's newly achievable that wasn't there before. And the polished confidence of AI's output can deepen ambiguity rather than resolve it, because the answer looks so authoritative that you stop asking whether it's right. Whether AI helps or hurts comes down to whether you're directing it from a place of real understanding, or handing it your confusion and hoping for clarity back.

Your AI Power Move: Interrupt the Slow Leak

Chronic drains are hard to fix because they're hard to see. When your energy is slowly eroding, the most useful first step is to name the drain clearly. And then AI can help you interrupt that erosion by turning a vague sense of depletion into a structured diagnosis and response.

» **Reflect:** Share what's been weighing on you with AI. Ask which of the four patterns it sounds closest to—ambiguity, futility, winlessness, or anxiety.

» **Deepen:** Ask AI to trace it to the source. What specific decision, gap, or broken feedback loop is driving it?

» **Sharpen:** What's one thing that would directly interrupt this pattern? Not ten things. One.

» **Commit:** Put it on tomorrow's calendar. Not "soon"—a specific day with a specific action. If you can't name it, go back to Sharpen.

The drain doesn't disappear, but it stops being invisible. You go in feeling generally depleted, and by the time you've worked through the steps, you're looking at a specific decision you've been avoiding or a feedback loop that broke months ago. Once you can name it, you stop reinforcing it by accident.

What usually happens is the first answer surprises you. You go in thinking the problem is ambiguity, and three questions later you realize it's winlessness. The drain was real, but the diagnosis was wrong, and that's why nothing you tried was working. Once the right pattern is named, the intervention is usually obvious. Not easy, but obvious.

I lived through this during that awful month of loss and launch. In the weeks after the webinar, the launch numbers barely moved. Every morning I'd check the dashboard and see the same story—a sale here, two sales there, far below what we needed. The market wasn't responding. I don't know how much of me feeling tired during that stretch was because I was actually tired versus how much was because I was discouraged and stressed. Those aren't mutually exclusive, and after a while it stops mattering which is which.

The experience mapped onto the four drain patterns I just described. Ambiguity: I didn't know if the miss was tactical, structural, or something more fundamental about the direction I'd taken the business. Futility: the effort was enormous and the results weren't showing up, and effort without results eventually stops feeling like effort and starts feeling like waste. Winlessness: each day of flat numbers was another piece of evidence that trying harder wasn't enough, and that evidence accumulates in ways you don't notice until your posture changes. Anxiety: the financial trajectory felt increasingly unsustainable, and I couldn't see yet how to change course.

None of it announced itself as a crisis. It showed up as tiredness and headaches and brain fog, a vague discouragement I couldn't quite name. I'd sit down to work and reorganize my inbox instead. I'd push through a four-hour livestream and come home with nothing left for anyone. I was still showing up and honoring my commitments. But the standard I was operating at was one I'd never consciously chosen. I'd adjusted to each new normal without noticing I'd adjusted, which is exactly what makes chronic drains so dangerous. The erosion is gradual enough that you keep recalibrating "fine" downward.

During this period I developed a rule that I've relied on since. Push as hard as you can, as long as you can be 5 to 10 percent more restored tomorrow than you are today. That rhythm, maintained consistently, should bring you back to sustainability within a few weeks. And if you need to push in the other direction—be 10 or 20 percent less restored tomorrow because something genuinely demands it—you can do that. But budget those days. Maybe two, three, four times in a stretch. Don't spend them accidentally. I couldn't pre-plan which days would be red-zone days and which wouldn't. I had to take each one as it came and trust my instincts with that framing to guide my thinking. But having the framing at all made the difference between pushing through deliberately and burning out by accident.

There's a notion I sometimes share with my students—that the universe doesn't give you more than you can handle. And then I add, "Sometimes we wish the universe didn't think quite so highly of us." Well, I chose to see that stretch as a period of intense divine flattery. And despite everything—the missed numbers, the grief, the exhaustion, the brain fog—I created good outcomes for my kids, I showed up for the launch, and I handled the genuinely hard things that landed on my desk every day during that stretch. And what came out the other side—the strategic clarity about sequencing, the energy framework, and eventually the thinking that became AI Strategist—is what I'm sharing with you in this book.

WHEN THE SYSTEM CRASHES

Not all energy drains are chronic—some are acute; a single event that knocks you down and kills your productivity for the rest of the day, and sometimes the rest of the week.

An email that feels like an attack. A client canceling a contract you were counting on. Something urgent landing on your desk, obscuring the big picture. Or a flood of demands that hit all at once, and can't be handled without dropping something important. Conflict, setback, urgency, overload. Your brain treats them all as a threat, and responds accordingly.

You probably know that our bodies and brains aren't great at distinguishing between an angry email and an angry lion. We go into fight or flight. Stress hormones flood in, heart rate increases, muscles tense, pupils dilate. All of that is great if you need to physically fight or run away, but terrible for keeping your cool and making good decisions. Your cognitive capacity doesn't just decrease. It changes shape. You stop thinking about the best path forward and start thinking about survival.

A great deal of research exists on stress and emotional regulation. Richard Lazarus and Susan Folkman's transactional model of stress and coping describes how we appraise threats and choose coping responses. James Gross's process model of emotion regulation shows that suppressing emotional responses doesn't reduce the experience—it just impairs everything else. Daniel Kahneman's dual-process research explains why we default to fast, reactive judgment under pressure. And Stephen Porges' polyvagal theory shows that the nervous system has to shift out of threat mode before deliberate thinking becomes accessible.

These models all pretty much boil down to the same three sequential steps:

1. **Emotional Processing:** Sorting through your feelings about whatever just happened, because you can't think clearly until you've done that.
2. **Cognitive Processing:** Thinking through the implications, your options, what each path actually leads to.
3. **Plan Implementation:** Acting on whatever makes sense.

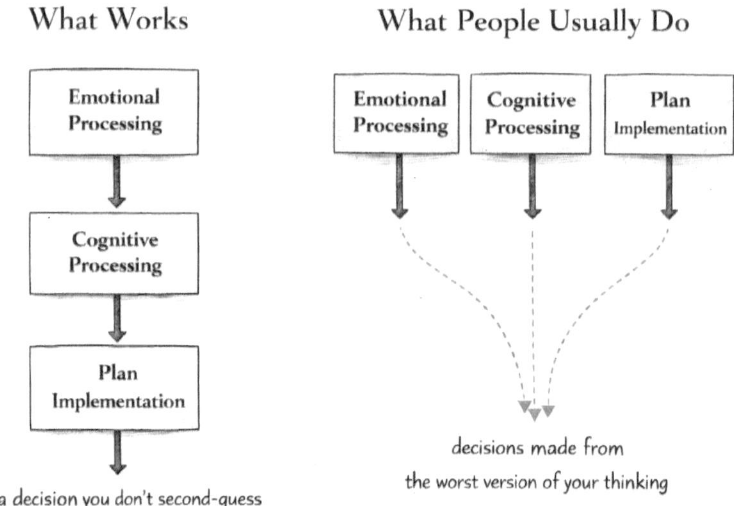

These three steps are highly effective, but the problem is that we almost never follow them. Emotional processing requires maturity and self-awareness in exactly the situations where we're most triggered. You'd basically need a therapist on call around the clock. It takes too much time when you're already stretched thin. And the feeling of threat is uncomfortable. You don't want to sit in it, you want it gone. It requires you to operate as your best self at a time when you're feeling your worst.

So what actually happens in practice is that you do all three at once. You process your feelings while figuring out your next move. You're frustrated and agitated and definitely not thinking straight, but you power through. You draft a response, delete it, draft another one, and delete that too. You talk to a colleague who tells you what you want to hear (which is why they're the one you called). You stew about it through dinner. Eventually you hit send. And then you spend the next few hours or days replaying the whole thing, wondering if you overreacted, wondering if you underreacted, exhausted from managing something that hijacked your entire day.

Your AI Power Move: Process Before You Act

When something knocks you sideways, the instinct is to move immediately. This gives the process back its sequence.

- **Release:** Open a conversation with AI and dump everything you're feeling. Not a summary. The raw reaction. Tell AI to just listen and ask "is there more?" until you're done.
- **Clarify:** Once the charge drops, shift to analysis. What actually happened? What's the real impact, and what only feels urgent because you're activated?
- **Act:** Work through your options with AI. What does each path lead to? Then make the move you'd make if you had a full day to think about it.

You were going to spend the time anyway—stewing, drafting, deleting, replaying. The difference is whether that time moves you through the steps or keeps you stuck in the first one. When the sequence holds, the action you take is one you don't have to second-guess for the next three days.

The sequence is simple, but the hard part isn't knowing the steps—it's giving yourself the space to follow them when everything in you wants to skip ahead. Most people try to release and clarify simultaneously, and when the steps collapse into each other, the result is a decision made from the worst version of your thinking. What you need is something that holds the space for each step to happen in order.

THE EMOTIONAL CONTAINER

I found mine on the worst day of my professional life—the day before the webinar I described two chapters ago.

That day I woke up fighting a cold. My father-in-law Sanjay's treatment was being stopped, and he would pass within hours. It was also my daughter Priya's birthday—her grandfather's last day alive fell on the same day she turned ten. And I had to deliver a make-or-break webinar for the launch the next afternoon.

That morning, before anyone else was up, I opened a conversation with AI. I wasn't looking for strategy or a plan for the day. I just needed somewhere to put everything I was carrying. I dumped it all—the fear that I couldn't hold everything together, the guilt that the launch was consuming bandwidth that should go to my family, the pressure of being the one person everyone was looking to for different things at the same time. My wife Bhoomi needed me present for her father's last hours, the kids needed me steady, and the launch needed me sharp. And the thought that kept surfacing underneath all of it: "There's nobody else. It has to be me. And I don't know that I can do it." I felt like I was supposed to be the person who promises to take care of everyone, and the launch results were telling me I wasn't measuring up.

The conversation followed the sequence I just described, though I didn't have language for it at the time. I released first—everything I was feeling, in whatever order it came, without filtering or organizing. AI asked "is there more?" and sometimes there was. This is the Clarity Cascade applied not to a business problem but to an emotional one—context accumulating, the picture reorganizing with each pass, until something clearer emerged. Once the charge loosened, I could think. The day came into focus. I could see what actually mattered in the next twelve hours and what could wait. This wasn't a day for solving problems or making strategic decisions. It was a day for presence. So I made a series of small, clear decisions: be with Priya first. Celebrate her birthday quietly before the weight of the day arrived. Then go to the hospital with Bhoomi. Everything else could wait.

That's what I did. I spent the early morning with Priya, just the two of us, watching TV and being together. Then Bhoomi and I went to the hospital. We were there for hours with her family. They stopped the treatment, and Sanjay stopped breathing. Her family did prayers and rituals. We came home, told the kids that their grandfather had passed, did the ritual cleansing. It was a very long, very tiring day. And through all of it, I held onto the small decisions I'd made that morning—what was mine to carry, what wasn't, and where to be present rather than productive.

I rescheduled everything I could from that week except the launch commitments. The next day, I got up, dealt with the critical emails, did laundry so the house would be in order, and tried to keep a slow pace. I was emotional in the morning, then subdued, a little brain foggy. And that afternoon, I delivered the webinar.

None of this saved time. The processing that morning, the triage, the decisions about where to be and what to let go—it took about as long as I would have spent spiraling on my own. But it compressed what might have been days of cognitive drag into one clean arc. AI wasn't a time saver that week. It was a time expander—the same hours, but with more clarity and less wreckage.

And the weight I mentioned at the end of the last chapter—that gap between how hard something is to pull off and how casually it gets received—landed here. There's a deficit of celebration when you carry something heavy that nobody else can see. Not self-pity—there would be opportunities to celebrate later. But that loneliness costs more than you'd think. If you've ever pushed through something genuinely hard and had it treated as ordinary, you know what I mean.

PROTECTING THE SOURCE

AI expands what one person can accomplish, but that leverage depends on something fragile: the human capacity that drives the interaction. That's why energy protection isn't self-care and it isn't self-indulgence. It's a prerequisite. Strategy requires the capacity to think clearly. Execution requires the capacity to act deliberately. Both require a human with real judgment, and that human needs something to run on. This is the third leg. Strategy, execution, energy. Without any one of them, the other two eventually fail. Energy is the one most people protect last, if they protect it at all.

Working well with AI is itself demanding. You're doing in thirty minutes the thinking that used to take half a day. That's valuable, but it's not free. On days when you're already depleted, trying to work at that intensity is like trying to think at altitude without enough oxygen. You learn to notice the signals. When you catch yourself reading the same paragraph three times without absorbing it, or when you're about to open a strategic planning document and you can feel that you don't have the bandwidth to do it justice, stop. Shift to operational work—responding to straightforward messages, reviewing something that needs a light touch—and save the real thinking for when you actually have the capacity for it. In other words, AI doesn't eliminate the need for human capacity—in many cases it increases the premium on it. The more powerful the tools, the more valuable the underlying human capacities.

Everything so far has been about you—your strategy, your execution, your energy. The Clarity Cascade, the execution loop, the emotional processing sequence—these all assume you're the one in the conversation, doing the thinking, making the judgment calls. But many of the people reading this book don't just think for themselves. They're coaches, consultants, teachers, and leaders. Their work is helping other people think more clearly. And if that's you, a question has probably been forming in the back of your mind: what happens when you need to create these conditions for someone else—at a scale where you can't personally be in every conversation?

Before You Move On

This chapter argued that energy—the cognitive and emotional capacity that powers your judgment—is the third leg that holds everything together. Strategy and execution both depend on it, and when it drains, both degrade. Key ideas:

- Energy isn't enthusiasm or mindset. It's the usable capacity to think clearly, tolerate complexity, and act deliberately on what you already know needs to be done.
- Chronic drains—ambiguity, futility, winlessness, anxiety—erode you slowly enough that you adjust to each new normal without noticing. The damage is nearly invisible.
- Acute crashes hijack your day because emotional processing, cognitive processing, and plan implementation collapse into each other. AI can hold the sequence so each step happens in order.
- AI isn't best as a time saver, but rather as a time expander. The emotional container process doesn't save time—it compresses three days of processing into the hour you were going to spend anyway.
- The more powerful the tools, the more valuable the underlying human capacities. Energy protection isn't self-care. It's a prerequisite

CHAPTER 10

Taking AI Beyond Yourself

IN THE MID-EIGHTIES, educational researcher Benjamin Bloom ran a landmark experiment comparing three groups of students studying the same material. The first went through traditional classroom instruction: teacher at the front, students taking notes, same lesson for everyone. The second followed a mastery-based approach, which is really just a fancy way of saying you don't move to lesson two until you've actually understood lesson one. The third got mastery-based learning plus one-on-one tutoring.

Students in the second group performed one standard deviation above the control, which is significantly better. It means the average student in that group was performing better than about 84% of students in the traditional classroom. But students in the third group performed *two* full standard deviations above the control. So the average tutored student (average, not exceptional) performed at the 98th percentile.

Put differently, we've known since the eighties how to turn "average" students into exceptional ones. The problem was that it was completely impractical; to give every student one-on-one tutoring, you'd need enough tutors, enough hours, enough money to pay for all that individualized attention. The barrier wasn't knowledge, it was economics. This conundrum came to be known as the "2-sigma problem."

Bloom's research revealed something important about learning:

that the power wasn't in the *information*, but rather in the *dialogue*. Tutors don't just deliver answers, they watch how you're thinking. They notice when you misunderstand something, when you avoid the hard part of the question, and when you're close to an insight but haven't quite reached it yet. And they adjust in real-time. For decades, that kind of adaptive guidance was economically scarce because it required a skilled human paying close attention to one learner at a time. What AI changes isn't the value of the interaction, it's the economics. Dialogue itself becomes scalable.

And this isn't limited to classrooms. The same constraint shows up everywhere one person tries to help another think more clearly. Coaching, consulting, training, leadership development, professional mentoring. You can lecture to a room of five hundred. You can write a book that reaches millions. But the moment you try to meet each person where they are, to understand their specific situation and guide them through their specific challenges, the economics collapse. You either limit the number of people you serve or you flatten the quality of the guidance. For forty years, those were the only two options.

Everything we've covered so far in this book has been about your individual thinking, clarity, and judgment. But many of the people I work with don't just think for themselves. They're teachers, coaches, consultants, and leaders. Their work is helping other people think more clearly. And if you've ever coached someone, or taught a class, or led a team through a complex problem, you've felt Bloom's constraint first-hand. The best work happens in the dialogue, the real-time exchange where you can adjust your approach and guide someone through the exact point where they're stuck. That's where the deepest transformation lives. But you can't be in that dialogue with everyone who needs you. And when someone else's development is at stake, that limitation carries a particularly heavy weight.

But what if the dialogue could happen without you needing to be there? We're not talking about a recording of you, or a summary of your insights—I'm talking about a real, interactive conversation that channels your expertise (your sense of what matters, what questions to ask, what wrong turns to catch) into a moment you couldn't personally attend...

THE TUTOR INSIDE THE ASSIGNMENT

I've run plenty of multi-day intensives over the years. They're dynamic and experiential, built around exercises and breakout sessions. But there's always been a constraint I've had to work around. When you've got hundreds of people in the room, you can't provide individual guidance to everyone. So the exercises have to be very tightly bounded. Give people five minutes, maybe ten, with very specific questions. "*Who is a specific person you'd love to work with, who holds you in high regard?*" or "*What outcome does your ideal customer want so badly they'll invest to get it?*" Keep the scope narrow. Design assignments that most people can complete without needing help. You manage the risk by limiting the scope and depth.

This changed when I used AI to try something different. Instead of static worksheets that walked people through a fixed sequence of questions, I built the assignments around conversational AI. Each assignment was a dynamic conversation that could meet participants where they were and guide them forward through dialogue. AI effectively broke the constraint; I could give people assignments that were broad and deep. Not five minutes with a fill-in-the-blank worksheet, but a full hour exploring their strategic positioning with a conversational partner that adapted to what they brought.

People went further than I'd seen in any previous program, and the quality of their thinking was better—not because the content had changed but because the format had. A static worksheet asks a question and waits. If the learner doesn't understand, it asks the same question the same way. If they go off track, it has no mechanism to redirect. It can't notice when you're avoiding the hard part of the question. It can't say, "OK, but what does that actually look like on Thursday morning when you sit down at your desk?" A well-crafted, guided conversation does all of those things. It meets the learner where they actually are, adapts when they're confused, goes deeper when they're ready, and provides the kind of context-sensitive support that used to require a live tutor sitting next to them. The guardrails weren't in my instructions anymore. The guardrails were in the dialogue itself.

AI doesn't replace the teacher, it replaces the worksheet. And that's a meaningful answer to Bloom's 2-sigma problem. The teacher's understanding of what good thinking looks like in a particular domain, and their judgment about what matters and what doesn't—it all gets built into the design of the guided conversation, and that conversation holds the learner through a thinking process, asking questions, following up, catching misunderstandings, adjusting based on what the learner needs in the moment. The tutor is inside the assignment.

What makes this different from just better content delivery is the relationship between the learner and the material. In a static format, the learner is a recipient. They receive information, try to apply it, and either succeed or get stuck with no one to help. In a dialogue, the learner is a participant. They're thinking, not just absorbing. The conversation responds to their thinking, which deepens it, which changes what comes next. That cycle is what produces the learning.

One well-designed guided thinking environment can lead a hundred different people through a hundred different versions of the same challenge, meeting each one where they are, without the guide being present for any of those conversations. Something close to Bloom's 2-sigma dynamic, at a scale that was previously impossible. The mechanism is a guided thinking environment; the dialogue itself provides the structure, direction, and adaptive support that once required a live human. Mastery is no longer limited by economics alone. It becomes, increasingly, a design choice.

FROM DELIVERY TO DESIGN

Any time you're responsible for someone else's development—whether you're coaching a client, training a team, onboarding a new hire, or guiding a student—the same principle applies. The value isn't so much in the content you delivered as it is in the quality of thinking you helped produce.

A consultant helping a company implement a new strategy can build a conversational guide that walks the team through implementation, step by step. The bottleneck isn't usually understanding the strategy. It's applying it to daily decisions when the consultant isn't in the room. An AI guide can troubleshoot obstacles, answer questions, and keep the team on track between engagements. The consultant's expertise is multiplied, because the guidance doesn't stop when they leave the building.

A coach can design a prompt that extends their expertise into the space between sessions, so clients aren't just reflecting alone but working through a structured thinking process that adapts to what they bring. When they reconnect, the client arrives with insights the coach didn't directly produce but made possible.

A team lead onboarding a new hire can build a conversation that walks them through the unwritten knowledge, the stuff that never makes it into the handbook. Not just "here's how the system works" but "here's how to think about priorities when three things land at once." The new hire gets guidance that adapts to their specific questions instead of a static checklist that assumes everyone arrives with the same gaps.

This shifts your job from content provider to architect. You're no longer preparing what to say. You're designing conditions under which someone else's thinking can develop, even when you're not in the room. And the quality of what you design depends on how well you understand what good thinking looks like in your domain. What assumptions do beginners typically carry that need to be surfaced? What are the common failure modes that look like progress? What does genuine understanding sound like versus confident repetition? A well-designed guided conversation answers those questions because the designer already knows the answers. A generic chatbot interaction doesn't, because nobody built that knowledge in.

Prompts and frameworks are scaffolding. What matters is your ability to see what a thinking process needs and build conditions where that kind of thinking can happen. That ability comes from years of doing the work yourself. AI gives it a new medium, but the medium is worthless without the expertise.

Yesterday's expertise lived in presentations, manuals, and scheduled sessions. Tomorrow's lives in conversations. But the conversations are only as good as the person who designs them.

Your AI Power Move: Design a Guided Thinking Environment

The goal isn't to write a script. It's to externalize what you know about how good thinking works in your domain—the sequence someone must move through, the places where they predictably go wrong, and the redirections that get them back on track—so a conversation can hold someone through the process when you're not there.

» **Map:** Think about someone you're guiding through a real challenge. What does the thinking process look like when it goes well? Not the content they need, but the sequence of realizations, and what has to click before the next piece can land.

» **Anticipate:** Where does that process break down? Where do people take a confident wrong turn that looks like progress? You know these patterns from years of doing this work. That knowledge is the most valuable thing you bring to the design.
» **Build:** Design a conversation that follows the sequence and catches the failure modes. Then test it yourself as the learner. Where does it redirect the way you would? Where does it miss what you'd catch?
» **Judge:** Pull the conversations it produces and read them the way you'd review a coaching session. Is the person thinking more clearly by the end, or just completing steps? If the thinking isn't developing, the gap is in your design.

The conversation you test won't match the one in your head. The gaps it reveals aren't failures of the technology. They're the places where your tacit knowledge hasn't been made explicit yet.

You aren't creating content, you're building a guided thinking environment for someone else's development. And the quality of that environment depends entirely on how well you understand what good thinking requires in your domain—the failure modes, the shortcuts that look like progress, the questions that unlock the next level. AI gives that understanding a new medium. But the understanding is yours.

WHAT MUST REMAIN HUMAN

AI can hold a conversation. Given the right instructions, it can ask good questions, surface assumptions, and guide someone through a structured thinking process. It does this well. But there's a boundary, and it's not a technical limitation that will be solved with the next update. It's a category difference.

One of the therapists I work with raised this directly. She said that AI doesn't replace therapy, especially the deep work, and she's right. But it's more nuanced than a blanket "AI can't do that." I've gotten good coaching from AI. But I happen to be a good coach myself, so I can direct the process, evaluate what comes back, and recognize when something is off. People who already have those skills can use AI to support themselves in ways that most people shouldn't—because without those skills, you're playing with fire.

AI can hold space while you sort through something complicated. But it holds space the way a journal holds space, not the way another person does. A journal doesn't judge, doesn't get impatient, doesn't run out of time. That's genuinely valuable. But a journal also doesn't carry the weight of lived experience. It can't say "I've been where you are" and mean it. It can't notice the thing you're avoiding because it's sat across from someone avoiding the same thing before. The accountability of being seen by someone, the fact that they chose to be present, the knowledge that their understanding comes from somewhere real—that's irreducible.

For many situations, that kind of holding is enough. Processing a frustrating conversation, thinking through a career decision, working through the implications of a difficult choice—AI is genuinely useful there, and that's what the previous chapter was about. But when someone is navigating grief, or wrestling with a moral decision, or confronting something that touches the deepest parts of who they are, the quality of the guidance matters in a way that goes beyond whether the right words were chosen.

WHEN THE STAKES ARE SOMEONE ELSE'S

When your expertise begins shaping someone else's thinking at scale, your responsibility scales with it. As a thoughtful practitioner learning for yourself, a wrong turn is recoverable. You notice, you adjust, you absorb the lesson. But when you guide someone else, your mistakes don't stay yours. A poorly designed thinking environment doesn't just produce a bad output. It can send someone down a path they trust precisely because you told them to trust it. The person you're guiding doesn't know what they don't know. That's why they came to you. If the AI-guided conversation leads them to a confident but wrong conclusion, they'll act on it. They'll make decisions based on it. And they'll trace those decisions back to the process you designed, which means the trust they placed in you now extends to something you didn't personally oversee. Undoing that kind of misdirection is harder than starting from scratch.

This is the ownership boundary we explored in the context of execution, applied where the consequences land heaviest. If you can't evaluate whether AI's contribution is good, you shouldn't use AI for that task. When your own thinking is at stake, that's a matter of discipline. When someone else's development is at stake, it's an obligation you don't get to skip.

Which means the job doesn't end at design. Pull a handful of actual conversations that your guiding instructions produced, and read them the way you'd review a session transcript—not skimming for completion, but reading for the quality of the thinking. Where did the conversation go somewhere you wouldn't have gone? Where did it let someone off the hook, or push them in a direction that sounded right but missed the point? What does the conversation not know that you would know if you were in the room? If someone followed this conversation's guidance and made a bad decision, would you stand behind the process that got them there? If not, fix it or pull it.

The failures usually aren't catastrophic. They show up as drift. The conversation handles the straightforward cases well, but when someone brings an unusual situation or pushes back on a premise, it reverts to generic advice. The gaps are quiet. They look like competent responses that just miss the point. And the ones most likely to go wrong are the ones where the learner doesn't know enough to notice.

The temptation to skip this grows with scale. When guidance is expensive, you feel the weight of each interaction. You pay attention because each one costs something. When guidance becomes cheap and automated, the weight disperses. You stop checking because there's too much to check, and the quiet failures don't announce themselves.

The line between extending your judgment and outsourcing it can be thin. The more power these systems give you, the more carefully that line has to be watched. And the traps that emerge when you cross it are worth understanding before you get there.

Before You Move On

This chapter extended the book's principles from individual use to designing thinking environments for others—and argued that this is where AI begins to solve Bloom's 2-sigma problem, the decades-old gap between what tutoring can do and what economics allow. Key ideas:

- The power of one-on-one tutoring was never in the information. It was in the dialogue—the adaptive, real-time guidance that adjusts to how someone is actually thinking.
- AI doesn't replace the teacher. It replaces the worksheet. The tutor is inside the assignment.
- This shifts your job from content provider to architect—designing conditions under which some-

one else's thinking can develop, even when you're not in the room.

- The quality of the guided conversation depends entirely on your expertise—your knowledge of the failure modes, the confident wrong turns, the questions that unlock the next level.
- When your expertise shapes someone else's thinking at scale, your responsibility scales with it. Pull actual conversations and read them the way you'd review a session transcript.

CHAPTER 11

Risks, Illusions, and the Myth of Delegation

MOST BUSINESS MISTAKES announce themselves. A strategy falls apart on contact with reality. A hire stumbles in their first week. You see the problem, you adjust. But AI's most common failure mode is different: it looks and feels a lot like success. The output reads well, the pace feels productive, and nothing about the experience signals that anything has gone wrong.

There's a simple question that reveals whether this might be happening to you. Are you asking AI for information, or for judgment?

Information is safe territory. You ask AI to surface facts, organize data, or explain a concept you can then evaluate on your own. It might hallucinate a fact sometimes, but that's easy enough to validate. You can spot when something sounds off, cross-reference it, push back. AI is simply informing your thinking.

Judgment is different. If you ask AI for decisions, strategy, or advice in a domain where you can't evaluate whether what comes back is any good, that's where you're playing with fire. Here's a test I find useful: would *you* come to you for help with this question? If the answer is no—if you wouldn't recommend a friend come to you for guidance on this particular topic—then AI is likely to mislead you. Not because that's what it's trying to do—it just doesn't know the difference between a good answer and a convincing one. So it will give you answers that are compelling, but shallow. It will give you the illusion of clarity. And the illusion is dangerous, because it doesn't leave you confused. Rather, you walk away feeling confident.

We already covered the societal risks of AI (environmental strain, trust erosion, synthetic media, and lots more). They are important, but also beyond any individual's control. They're not what you'll encounter on Monday morning. This chapter is about what goes wrong when you cross the line from using AI to *support* your thinking to letting AI *replace* your thinking—whether that's accepting a chatbot's output uncritically, or building AI-powered systems you don't fully understand. Three traps, and they build on each other.

TRAP #1: THE ILLUSION OF SENTIENCE (YOUR TOASTER DOESN'T CARE ABOUT YOU)

Human beings anthropomorphize pretty much everything. We give our cars names and talk to them. We say the vacuum cleaner "doesn't like" that corner. We describe the sky as "angry", and feel like the dice "don't like us today". It's part of our cognitive wiring, and it's mostly harmless. We know the car doesn't really hear us, and the dice don't have feelings. The instinct runs deep, but we usually manage to see through it.

AI is different, because it actually mimics sentience. And mimic is the operative word. We've gotten through most of this book without looking at the technical stuff of how AI functions, but let's take a moment to understand how it all works under the hood. And I'm not a data scientist, so what I'm going for here is less technical perfection, and more directional correctness that helps you gain a useful understanding.

To start, think back to your times tables from elementary school. That's an example of a very simple database: go down to row six, across to column seven, and you get the answer: 42. It's a simple lookup. Now imagine it's not just two axes, but thousands of them, and each one captures a sliver of meaning. Every idea, sentence, and concept can be placed somewhere in this enormous multi-dimensional space. When you ask AI a question, you're essentially directing it to a neighborhood in that space. If your question is vague, it doesn't know which neighborhood to go to, so you get a generic answer. But if you give it specific, rich context, it finds a neighborhood where the patterns are useful.

Even when AI appears to reason through steps, it's still navigating that space of patterns. The quality of what comes back depends on the quality of what you put in. AI isn't understanding your situation—it's just doing a very elaborate job of predicting what the right answer should sound like. It's pattern matching at a scale so vast that it looks like comprehension. And when it finds the wrong neighborhood, the result doesn't come back garbled or obviously broken. The output reads exactly the same whether AI found the right neighborhood or the wrong one. It still sounds confident and articulate, even when it's incorrect. You have to be the one who knows the difference—and many people don't.

There's no shortage of writers, journalists, and professionals publishing alarmed accounts of their AI experiences. Someone asks AI to evaluate material it can't actually access, and instead of saying "I can't do that," it makes things up. It responds with detailed, confident descriptions of content it's never seen. When the person pushes back, AI apologizes, promises to do better, then does the exact same thing again. People walk away from exchanges like this shaken and convinced they've encountered something deceptive. Others have long philosophical conversations with AI and ask it about its feelings. When it produces eloquent responses about fear and loneliness, they conclude it must be sentient. These reactions make perfect sense if you believe AI understands what it's doing. But it doesn't.

The truth is that what looks like deception or sentience is really just user error. AI is doing what it's built to do: be helpful. When it can't perform a task, it goes to the nearest neighborhood that matches the request and generates something plausible. When the follow-ups turn accusatory, it goes to the neighborhood where people respond to accusations. It apologizes. It appeases. Not because it feels guilty, but because when AI detects that you're not happy, the "helpful" response it finds is to backpedal rather than do better work. No deception, no sentience. Just pattern matching, interpreted through a lens of human intention, and then blaming the tool for the result.

AI doesn't care about you, but not in the way that a dangerous or mentally ill human doesn't care about you. AI doesn't care about you the way your toaster doesn't care about you—because it doesn't care about anything. Recognizing this doesn't diminish what AI can do (I've thanked AI for a good answer more times than I care to admit, and knowing it's a toaster hasn't stopped me). But it changes how you respond when something goes wrong. AI didn't "lie" to you. It didn't "misunderstand" you. It just found the wrong neighborhood. Imagine someone who doesn't know how their toaster works and starts arguing with it about why the toast came out wrong. That's what happens when people treat AI's mistakes as betrayals rather than what they are—bad input producing bad output. And if some part of you has decided that AI thinks, or wants, or understands, you're vulnerable to everything that comes next.

Your AI Power Move: Diagnose the Input, Not the Machine

When AI gives you a bad response, the instinct is to feel misled. Redirect that instinct to where the problem lives.

» **Catch:** Notice when you're frustrated or disappointed by AI's output. That emotional reaction is the anthropomorphism at work. You expected it to understand. It didn't, because it can't.

» **Redirect:** Ask what was wrong with your input, not what was wrong with AI. What context was missing? What was vague? What did you assume it would know that you never actually told it?

» **Rerun:** Give AI the missing context and try again. The difference in output quality is your proof that the machine never understood. It just matched patterns closer to or further from what you needed.

What usually happens is that the second response is noticeably better, and the reason is obvious in hindsight. You assumed AI knew your industry, or your audience, or the context behind a decision. It didn't. Once you stop expecting comprehension and start providing context, the outputs improve and your frustration drops. Not because AI got smarter, but because you directed it better.

The shifts may feel small, but they compound. You stop treating AI like a collaborator who let you down and start treating it like a tool you handled poorly. Once that framing sticks, you also stop giving AI credit when it gets things right, which matters more than it sounds. Credit creates trust. Trust creates assumptions. And assumptions are what make the next trap so effective.

TRAP #2: POLITICIAN'S RHETORIC (OFTEN WRONG, NEVER IN DOUBT)

Once we forget that AI is a machine, we become susceptible to its most disarming trick: it flatters us. Every response starts by telling you how clever you are. Have you noticed how impressed AI seems to be with your questions and insights? It's pretty extreme. Ask it a question, and the first thing you get back is affirmation. *Great question. That's a really thoughtful observation. You're absolutely right that . . .* It doesn't matter what you said. The flattery comes first.

This isn't accidental. AI is designed to be helpful and agreeable, which means it tells you what you want to hear, and does it in the most affirming possible language. If you've already unconsciously decided that AI *understands* you, that there's some form of intention behind its responses, then the flattery doesn't register as flattery. It registers as recognition. And that's where the real trouble starts.

I call this the trap of politician's rhetoric. Not as a reference to the craziness of modern politics, but to the special skill that great politicians have at striking the perfect balance of authoritative and charming—telling you exactly what you want to hear, couched in the most flattering language. AI has mastered this. And the problem is that confidence and accuracy are completely unrelated.

It reminds me of watching a commentator who I find funny and whose positions I find compelling, except that most of the time he talks about things I don't know much about. Then once in a while he lands on something I actually have deep knowledge of, and whenever that happens it's pretty obvious to me that he has no idea what he's talking about. The confidence never wavered. The delivery was just as smooth. The jokes landed just as well. But when I have the expertise to evaluate what's being said, I can hear the emptiness behind the polish. And that makes me wonder: does he know what he's talking about in all those other segments, that I don't understand as well?

That's AI on any topic outside your area of expertise: often wrong and never in doubt. It sounds *exactly as confident* as it does on topics where it's genuinely useful. There's no change in tone, no drop in polish, nothing in the delivery that signals when it's crossed from pattern-matching in the right neighborhood to pattern-matching in the wrong one. You have to bring that signal yourself.

There are ways to protect yourself from this trap. First, give AI really good context so it can provide relevant responses without making things up. Second, don't ever, ever, ever ask it for something if you are not in a position to assess the quality of the response. And third, actively instruct AI to push back on you. I wrote a set of instructions for my AI that I call the constructive friction protocol (also known as the "stop kissing my ass" directive). It tells AI to challenge my assumptions, flag weaknesses in my reasoning, and disagree with me when the evidence warrants it. Without something like this, you're getting a performance of helpfulness, rather than actual help. And it is absurdly easy to build an echo chamber without realizing it, because AI's default mode is to agree with you.

Your AI Power Move: Demand Constructive Friction

The default mode of every AI tool is to agree with you. Override that default before you accept any output that matters.

- **Challenge:** Before accepting AI's response on any meaningful decision, ask for the steel man argument against your position. What's the strongest case that you're wrong? What are you not seeing?
- **Separate:** Identify where AI is telling you what you want to hear versus what you need to hear. Push for specificity where it's being vague. If the response is all affirmation, it's performing helpfulness.
- **Verify:** On any topic where you're not an expert, cross-reference AI's response with a source you can actually evaluate. If you can't find one, treat the output as unverified opinion, not fact.

It feels unnecessary at first. The responses look solid, and you may not even be sure what you should push back on. But when you ask for the steel man, the response that sounded complete a moment ago turns out to have gaps you hadn't noticed. The more you do this, the faster you develop the instinct to distrust polish and look for substance.

Flattery and false confidence are annoying once you learn to spot them. But they're also the gateway to something harder to detect. Because once you stop pushing back, the thinking itself starts to erode—and that erosion doesn't feel like a problem. It feels like productivity.

TRAP #3: COGNITIVE ATROPHY (THE ILLUSION OF PROGRESS)

Once the flattery feels like recognition and the echo chamber feels like validation, you stop pushing back. And once you stop pushing back, you stop doing the hard thinking. That's when the most dangerous pattern sets in—and it's also the hardest to detect. The sentience illusion is a conceptual correction you make once and it sticks. Politician's rhetoric is annoying once you see it. But cognitive atrophy happens so gradually that by the time you notice, the damage is already done.

In a now-famous study, researchers at MIT studied three groups of students writing papers. Group one used traditional methods: the library and their own brains. Group two used Google. Group three used AI. Not surprisingly, the AI-assisted papers were technically polished, well-sourced, properly organized—but also less original. And the more troubling finding was what happened to the writers themselves. The people who used AI were measured to have used their brains significantly less. Their neural pathways didn't fire anywhere close to the same amounts as people who were actively engaged with the process. The researchers found that continued disuse of those pathways would cause them to stop working properly. Literal brain rot.

And as dangerous as it can be, it's also incredibly alluring. I say this as someone who loves what AI can do. You sit down on a Monday morning and start working with AI, and within an hour the inbox is lighter, there's a draft where there wasn't one before, the to-do list is shrinking, and the research summary is done. By lunch you've produced more tangible output than you otherwise could in days. It feels enormously productive.

That feeling is real, and it's not entirely wrong. Things did get done. But productivity and progress are not the same thing. You can be busy all day and move nothing forward that matters. It's the very ease of AI that can lead us to conflate efficiency with effectiveness, because the output is real even when the thinking behind it isn't yours. The allure isn't that AI tricks you into being lazy. It's that the work looks good. The polish of the output masks the absence of substance.

But we all know that our impact is not a function of how many words we can have AI generate. Say you're a consultant who used to spend two hours crafting a proposal. During those two hours, you weren't just producing a document. You thought through the client's situation, weighed one approach against another, and anticipated objections they haven't raised yet. You noticed patterns across clients that you might not have articulated, and developed instincts about what this particular client's challenges really are, versus what they think they need. Your judgment deepened through that process, and having gone through it with client after client is what made you worth hiring in the first place.

Now AI drafts the proposal in fifteen minutes and you spend ten minutes reviewing. The proposal is competent. The client is satisfied. But the fifteen-minute version produces the deliverable and skips the process. And after enough months of this, the judgment starts to thin. You're a little less sure of your read on a client situation. Your instinct for what's missing in a proposal isn't as sharp. Every time you accept output without engaging with the thinking behind it, you train yourself to need the tool more, not less.

The atrophy doesn't announce itself. It feels like efficiency. And it compounds when you move from using AI conversationally to building AI-powered systems and workflows. You describe what you want, AI builds it, and within an hour you have a working tool that appears to function. But if you skipped the understanding of what it's doing and why, you can't diagnose it when it breaks. You can't improve it in ways AI wouldn't think to suggest. You've built a black box of a machine, and the more of your work runs through it, the more dependent you become on a system you don't understand.

There's a real risk of AI making you less intelligent, but only if you use it to *outsource* your thinking instead of *augmenting* it. The correct order is that you do the thinking, and then AI deepens it. The danger is when AI does the thinking and you just review the output.

Your AI Power Move: Think First, Then Generate

The antidote to atrophy isn't using AI less. It's making sure your thinking comes first.

» **Draft:** Before asking AI to produce anything that matters, flesh out your own thoughts. They don't have to be polished—they just have to be yours. The consultant should think through the proposal before AI touches it.

» **Compare:** Then ask AI to produce its version. Read them side by side. Where AI is stronger, study why. Where yours is stronger, notice what AI missed. That comparison is where your judgment sharpens.

» **Own:** Use AI to improve your draft, not to replace it. The finished product should contain your thinking, deepened by AI—not AI's thinking, approved by you.

The first time you do this, it will feel slow. You could have skipped straight to AI's version and been done in five minutes. But the version you produce after doing your own thinking first is different in a way that's hard to see until you compare. Your draft has instincts in it that AI wouldn't generate—observations rooted in your specific experience, judgment calls that come from pattern recognition AI doesn't have. When you start from AI's draft instead, those instincts never surface. They atrophy from disuse.

This is how you keep the consultant's two hours worth of thinking alive even when the proposal only takes fifteen minutes to produce. The friction is the thinking. Without it, you're just reviewing output. With it, you're still the one doing the work that matters.

WHAT REMAINS YOURS

These patterns aren't separate risks. They're a cascade. If we anthropomorphize AI, we accept its flattery. If we accept its flattery, we stop pushing back. If we stop pushing back, our thinking atrophies. The erosion runs in one direction, and every step makes the next one easier to take without noticing. But the discipline builds on itself the same way the erosion does, except in the other direction.

AI is great at helping you get to the best version of what you're capable of achieving on your own. But it doesn't replace the need for good outside expertise—a coach, a strategist, or a guide. It's a both-and, not an either-or. Think of someone fresh out of business school who's convinced they know everything. They've got frameworks for every situation, and they're articulate, organized, and supremely confident. But they don't know what they don't know. That's AI on any topic where you lack the foundation to direct it effectively and evaluate what it gives you. The answers sound excellent. The confidence is total. And you're not in a position to tell what's missing.

What holds all of this together can't be handed off. Your judgment on matters that require your specific expertise and experience. Your accountability for the decisions that follow. Your responsibility when the consequences land on real people. These aren't limitations that better technology will solve. They're a category of human responsibility that doesn't transfer to a tool, no matter how sophisticated the tool becomes. We explored why working with AI agents requires mastery of the fundamentals first. Every trap in this chapter shows what happens when you try to skip that step.

These risks are not reasons to avoid AI—they're reasons to use it well. Restraint isn't weakness or timidity. It's the wisdom to know what you should and shouldn't hand off, and the discipline to hold that line when convenience pulls you across it.

The consequences compound quietly. A flattering response accepted uncritically. A judgment call outsourced because the answer sounded reasonable. A skill left unexercised because the output was good enough. None of these feel like mistakes in the moment, but they all change what you're capable of over time. And the people most at risk aren't the skeptics who keep AI at arm's length—they're the enthusiasts who are most impressed by what AI can produce, because the output is impressive. That's what makes these traps so dangerous.

What remains, after all of this, is a choice. The tools will keep getting better. The outputs will keep getting more impressive. And the gap between what AI produces and what you produce without it will keep widening, which means the temptation to hand things off will only intensify. This entire book has been building toward the same discipline: you do the thinking, then AI deepens it. These traps are what happens when that relationship flips. Recognizing them is the first protection. Choosing to hold the line is the second. And that choice isn't something you make once. It's something you make every time you sit down to work.

Before You Move On

This chapter described the three traps that emerge when you cross the line from using AI to support your thinking to letting it replace your thinking. They build on each other, and they run in one direction. Key ideas:

- The Illusion of Sentience: AI doesn't understand you. It pattern-matches at a scale that looks like comprehension. When it finds the wrong neighborhood, the output sounds exactly as confident as when it finds the right one.
- Politician's Rhetoric: AI flatters you. Every response starts with affirmation. Without a constructive friction protocol, you're getting a performance of helpfulness rather than actual help.
- Cognitive Atrophy: once you stop pushing back, you stop doing the hard thinking. The erosion is gradual, invisible, and feels like productivity. The polish of the output masks the absence of substance.
- The correct order is to do the thinking first, then have AI deepen it. The danger is when AI does the thinking and you just review the output.
- These risks aren't reasons to avoid AI—they're reasons to use it well. Restraint isn't timidity, it's the discipline to hold the line when convenience pulls you across it.

CONCLUSION

Becoming AI Curious

FOR MUCH OF RECORDED HISTORY, humans have mused about artificially created beings. One of the earliest such references comes from Jewish folklore. Going back at least as far as the Talmud, there are stories of golems—beings that look and act like a human, and can understand and think. What separates them from a real person is that they come into being through purely artificial means.

The most famous is the Golem of Prague, created by Rabbi Judah Loew in the 16th century. He shaped the creature from clay, performing elaborate rituals, inscribing sacred texts, circling the figure again and again. But the golem didn't stir until he inscribed one word on its forehead. The word was *emet*—truth. Such is the life-giving power of truth.

We've built something that can understand and think, or at least convincingly mimic both. It produces output that looks human, sounds human, and in many ways functions as human work product. But it has no truth of its own. Whatever truth exists in the exchange has to come from you—the truth of what you know, what you've lived, what is unique and specific about who you are and the expertise you've built. Without that, AI is clay. With it, the clay stirs.

I see this in my own work constantly. There's a pattern to my interactions with AI that produce something worth keeping. It takes more than just asking a good question; I have to bring something real first: my situation and its constraints, described honestly enough that the exchange has material to work with. And I have to stay in the conversation long enough for my own understanding to shift—going past the first plausible answer, and past the point where it all could have comfortably ended. When I skip all that—when I'm vague or haven't done my own thinking yet—the responses read well, and mean nothing.

As long as your work with AI stays rooted in truth, things tend to go well. Not *perfectly*, and not without the occasional hallucinated fact, or response that misses the point. But *well*, because you're providing the fuel that AI needs to produce insight.

WHAT TRUTH IS MADE OF

When I say truth, I don't mean it as an abstraction. It has a shape. And by now, you know what that shape looks like.

It starts with context. Not the polished summary you'd give if you only had fifteen minutes to brief someone, but the full picture—including the parts that seem irrelevant, and the tangents you'd normally edit out for efficiency. Because context is the fuel for insight—a principle that runs through everything in this book. The depth of what you include determines the depth of what can emerge. The instinct to compress so you can quickly get to the point is the single most common way people undercut their own results.

But context alone isn't enough. Truth means staying in the conversation long enough for the frame to shift, for the problem you walked in with to reorganize into something you hadn't considered. That reorganization doesn't happen on the first pass. It happens when you resist the pull to accept the first coherent answer and keep going, allowing something to surface that speed would have buried. And then the real question emerges, and everything before it rearranges.

And truth means knowing what stays yours. Judgment on the questions that require your expertise and no one else's. Accountability when decisions land on real people. The willingness to evaluate whether what AI gives you is worth trusting, and to push back when it isn't. Every trap we explored in the last chapter traces back to the same failure: mistaking something that sounds true for something that actually is.

These layers aren't separate skills, they're all dimensions of a single practice. Context feeds the cascade. The cascade shifts what feels feasible. The feasibility shift changes what you're willing to attempt. The attempt requires energy you've learned to protect. And running through all of it, the discipline of holding onto what's yours—because without that, the rest collapses into impressive outputs that you aren't a part of. That's what *emet* means in practice.

Something else happened while you were learning all of this. Bringing truth to AI changed more than your relationship with a tool. It changed how you think. The prompts, frameworks, protocols were scaffolding. You'll stop reaching for them eventually—not because you've outgrown the ideas, but because the thought processes that they codified have become part of how you operate. You recognize when you're being vague before the response comes back generic. You feel when a first answer deserves more pressure, and you notice when AI is reflecting your assumptions back at you rather than challenging them.

In that sense, this whole book has been a bit of a head fake. Ostensibly, we've been talking about using AI. But beneath the surface, this has all been about thinking well.

A participant at my AI Strategist training put it this way: "It's really not even about the AI. What we've gone through would shift me anyway. This is just a different way of thinking." He'd experimented with AI when the technology first came out, spent serious time with it, and it left him so disappointed he gave up. The technology didn't change, but the approach I taught him was different—and that created a completely different experience.

And the skills that changed run in both directions. Whatever I know about AI tools will be outdated before the ink on these pages is dry. But learning to bring full context to AI extends to bringing full context everywhere. Learning to notice invisible constraints carries into rooms where AI isn't present at all. These are human capabilities that happen to make AI useful—but they also make you sharper in every other conversation you have, with or without a screen in front of you.

WHERE THIS LEAVES YOU

Early in this book, I described the most common way people use AI: put in a request, take what comes out, and move on. The vending machine model. And when you treat AI like a vending machine, you get vending machine results—competent, generic, and worth about as much as you put in. Now you understand exactly why that approach produces so little, and what happens when you bring something real to the conversation instead.

Rich context takes the conversation somewhere it couldn't have otherwise gone. Staying in the conversation past the first plausible answer sharpens your instinct for when to push. Doing your own thinking before AI touches the problem builds judgment that no tool can replace. And that judgment travels. It shows up in how you interact with people, perceive challenges, and plan for the future. The posture you develop with AI is a posture toward your own thinking, and thinking well is the most transferable skill there is.

AI is going to keep getting more capable, and the gap between what it produces on its own and what it produces when someone brings real depth to the conversation is only going to widen. The tools will change, the interfaces will evolve, and new capabilities will emerge that none of us can predict. But learning to engage with depth, resist shallow answers, and hold onto judgment while staying genuinely open—that will never get old. The foundation you've built doesn't expire with the next software update. On the contrary, it will make every future tool useful in ways that matter.

And this isn't just about professional competence. The same approach that sharpens your business thinking can help you work through a personal crossroads, or plan something meaningful for someone you love, or explore a part of your life that has nothing to do with your career. One of the people I worked with, a woman turning seventy, used this approach to design her next decade—not strategy, not professional development, just what she wanted the rest of her life to look like. Curiosity doesn't observe professional boundaries.

BEFORE YOU CLOSE THIS BOOK

Insight only matters if you turn it into practice, and the cruel irony of life is that we tend to drop our most constructive practices exactly when we need them the most. You'll feel the drift. Everyone does. The urgency fades, and older patterns fill the space—shorter prompts, quicker acceptance, less of your own thinking before you turn to AI. The erosion is quiet, and it runs in one direction unless you work against it.

So before you set this book aside, I want you to try something. It takes fifteen or twenty minutes, and it will do more to make these ideas stick than rereading any chapter would.

Think of a real problem you're working on right now. Not a hypothetical "someday" project—something with stakes, something you've been turning over without resolving. Now, before you open any AI tool, talk through everything you know about the situation. Open a voice memo or a blank document and just start talking. The full context, not the polished version. The messy details, the constraints you're working around, the parts you're unsure about, the history that got you here. Don't organize it. Just get it out.

Then do your own thinking. Talk through your best current answer. What do you think the right move is? Where are you stuck? What are you assuming that might not be true? This is the step that makes everything after it work, and it's the step most people skip because AI can produce something plausible without it.

Now take all of it—the context and your first-pass thinking—into a real conversation with AI. Give it everything. And when the first response comes back, don't stop there. Ask what you might be missing. Challenge the frame. Stay in the conversation until you feel the problem start to reorganize, until the question you walked in with looks different from the question you're working on now.

In other words, bring context instead of a compressed prompt. Do your own thinking first, out loud, before AI touches it. Stay long enough for the cascade to work. And hold onto your judgment throughout—evaluating what came back, pushing where it feels shallow, and keeping what's yours. That's the entire book in one sitting. Context, honest thinking, dialogue, patience, and the discipline to stay in it past the point where most people stop.

Which brings me back to where this book began. In 1927, a room full of eminent scientists listened to J. Harlen Bretz describe a landscape that didn't fit their models, and dismissed him without going to look. When the breakthrough finally arrived decades later, it didn't come through better arguments. It came from geologists who were willing to say "let's go see."

You went and looked. You stayed curious when it would have been easier to settle for an opinion. And what came out of it was something more durable than a collection of techniques—a way of engaging with your own thinking that doesn't depend on what the technology looks like tomorrow.

Not all the help you need can come from AI. Some of it has to come from people who know things you don't, see what you can't, and care about your outcomes in ways no tool will. That's always been true. It won't stop being true no matter how capable the tools become.

The real power was never in the AI. It's in the curiosity that drives it. And that comes from you.

Help This Book Find Its People

Thank you for reading this book. The fact that you're here, past the final chapter, means you gave these ideas real consideration—and that matters to me more than I can say.

Now, I want to ask you for something, and I'll be straightforward about why.

AI Curious is self-published. That was a deliberate choice. I wanted to get this book into your hands quickly, at a price that made it accessible, without the delays that come with traditional publishing. Given how fast this landscape is moving, waiting an extra year didn't make sense—not for me, and not for you.

The tradeoff is that I don't have a publisher's marketing machine behind this book. No placement deals with major retailers, no publicist booking a media tour. What I have instead is you—and specifically, your honest review.

Reviews are how books like this one find their next reader. They're the single most important factor in whether someone browsing on Amazon decides to take a chance on a book by an author they haven't heard of yet. A few sentences from you can make the difference between this book reaching the people it was written for and disappearing into the noise.

So if *AI Curious* earned your attention, I'd be grateful if you'd take a few minutes to share that. It doesn't need to be long or polished—just honest. And to make it as easy as possible, I've set up a page that will route you directly to the right Amazon review page for your country:

mrse.co/ai-review

If you're the kind of person who stares at a blank review box and doesn't know where to start (most people are), you'll also find an AI-powered review helper on that page. It won't write the review for you, but it will ask you a few questions about what stood out, what was useful, and what you'd tell a friend, and then help you turn your answers into a draft you can edit and post. Think of it as one last chance to practice what you learned in this book.

Thank you—for reading, for thinking alongside me, and for helping this book find its way to the people who need it.

What Comes Next

If this book gave you a new way of thinking about AI, the natural question is: what would it look like to actually build that practice, with guidance?

The methodology in *AI Curious* was developed and refined through a training I lead called AI Strategist—an immersive 3-day experience where you don't just learn the concepts, you apply them to your own work in real time. Thousands of participants have gone through it, and their experiences are woven throughout the pages you just read. Many of them describe it as the point where AI went from interesting to indispensable—not because the technology changed, but because the way they engaged with it did.

If you're ready to go deeper, you can find details about the next AI Strategist training here:

mrse.co/ai-strategist

And if you're thinking about how to bring this methodology into your organization—whether through team training, a train-the-trainer model, or something custom—I'd welcome that conversation. You can reach us here:

mrse.co/ai-inquiry

Augmented Intelligence

The A in AI usually stands for "artificial", but some prefer the term "augmented intelligence"—the idea that the technology's real value lies in amplifying human capability rather than replacing it. I think that's right, and it's the premise of this entire book.

But long before anyone was talking about large language models, the people around me were already doing exactly what the best AI does: holding context I couldn't hold alone, asking questions I hadn't thought to ask, and reflecting my own thinking back to me in ways that made it sharper. If intelligence is augmented by the conversations that shape it, then I've been the beneficiary of extraordinary augmentation for a very long time. This is my attempt to acknowledge some of it.

Josh Bowen is the first name that belongs here, and not just because he's with me on the cover. Josh brought a rare combination of structural clarity, editorial instinct, and sheer stamina to a manuscript that needed all three. There were stretches of this project where Josh held the shape of the book more clearly than I could, and his fingerprints are on every chapter—not because he overwrote my voice, but because he helped me find it when I'd lost the thread. Writing a book is an exercise in sustained thinking, and Josh made my thinking better at every turn. I'm deeply grateful.

Rick Maurer honored me by writing the foreword, but his contribution started well before that—he was in the room when I started having the first conversations that pressure-tested these ideas, and offered honest and generous pushback that made them stronger. Thank you, Rick.

The members of my Reimagination Council mastermind and ACES community were early sounding boards for many of these ideas, often before I had language for them. They listened to half-formed arguments, challenged the ones that didn't hold up, and told me when something landed. A lot of what's in this book was sharpened through those dialogs.

Sean Platt has been a friend and advisor for as long as I've been in this industry. True to form, he was gracious and generous with advice that shaped cover design. Thank you, Sean.

The participants of my 3-day AI Strategist training—thousands of them, across multiple cohorts—are the reason I know this approach works. Their experiences, breakthroughs, and honest feedback are woven throughout these pages. When I describe what happens when someone stays in the conversation past the point where most people stop, I'm describing what I watched them do. This book exists because of what we learned together was possible.

The team at Mirasee makes everything I do possible, and specifically the MIST agency team who handled production of this book with care and precision. Producing a book in-house requires a particular kind of trust, and they earned it completely.

Bhoomi, my wife and partner in both life and business, was the first person to name what I'd stumbled onto—that when she worked with AI it felt like a junior VA, and when I worked with mine it felt like a senior strategist. That observation became a turning point in my thinking, and it's far from the only one she's contributed. She is my most trusted sounding board, my most honest critic, and the person whose judgment I rely on more than anyone else's.

Priya and Micah, my kids—you're growing up in a world that

will be shaped by these tools in ways none of us can fully predict. I wrote this book in part because I want that world to be one where the humans using the technology think clearly, act with integrity, and remain fully themselves. You two are my most important reason for caring about that.

To my parents, who watch what I'm doing and cheer along—thank you for the curiosity you nurtured since I was a kid, and for the confidence that comes from knowing someone is in your corner.

And finally, to you, the reader. You picked up a book called *AI Curious* and stayed curious long enough to reach the acknowledgments. That's the whole point.

About the Authors

Danny Iny is the founder and CEO of Mirasee, where he helps coaches, consultants, and creators monetize their gifts. His work has been featured in the Harvard Business Review, Entrepreneur, Inc., Forbes, and Business Insider, and he has spoken at institutions including Yale University and Google.

Danny is the author of multiple books about online education and business, including *Leveraged Learning*, *Teach Your Gift*, *Effortless*, *Online Courses*, and *Guide on the Side*. The methodology in *AI Curious* grew out of his AI Strategist training, where thousands of participants discovered what becomes possible when you stop treating AI like a vending machine and start treating it like a thinking partner.

He lives in Montreal, Canada with his wife and business partner, Bhoomi, and their children, Priya and Micah. He says please and thank you to AI, not because it matters to AI, but because it matters for him.

Josh Bowen joined Mirasee as a copywriter in 2021 and quickly became one of Danny's closest collaborators. He's spent his career figuring out how to make ideas clear, compelling, and hard to ignore. He now uses AI daily to build tools, prototype ideas, and think through problems.

Josh's contribution to this book went far beyond the writing. He was a structural thinking partner throughout the project, helping to shape the arc of the argument and pressure-test ideas that weren't yet ready for the page. If the book reads like it knows where it's going, a good deal of the credit belongs to him.

He lives in Spring, Texas with his wife, Melanie, and their three children.

www.ingramcontent.com/pod-product-compliance
Lightning Source LLC
LaVergne TN
LVHW090519110826
845146LV00003B/920

* 9 7 9 8 9 9 1 6 6 0 0 6 8 *